CHILD GROWTH AND EDUCATIONAL PSYCHOLOGY

CHILD GROWTH AND EDUCATIONAL PSYCHOLOGY

K.L. TULI

CYBER TECH PUBLICATIONS
4264/3, Ansari Road, Daryaganj, New Delhi-110002 (India)
Ph.: 011-23244078. Fax: 011-23280028
E-Mail: cyberpublicationsdelhi@yahoo.com

First Edition 2009

Published by G.S. Rawat
For Cyber Tech Publications
4264/3, Ansari Road, Daryaganj, New Delhi - 110 002
E-mail : cyberpublicationsdelhi@yahoo.com
Phone : 011-23244078,
Fax : 011-23280028

Printed at
Amit Enterprises, Moujpur, Delhi. Ph. - 9810878184

PREFACE

The Child Growth and Educational Psychology. When a child is born she is completely dependent on others for care and satisfaction of all her needs. The growth and development that takes place during early years and the care and protection provided by adults enable her to stand on her feet. The members of the family interact with her and over a period of time establish their own relationships.

Teaching the child the ways, beliefs, values and traditions of the family and society. In the process of bringing up their children parents teach them to acquire the ways, traditions, beliefs and values of the family and thus socialize them. The teach the children to respect all who are older to them, not to fight or be rude or aggressive, to be honest, take care of their things, to remove footwear when visiting a religious place or while attending a religious ceremony and to respect all religions and so on.

The aim of this book is very simply stated. The working teachers and students who have opted to study education as a course subject in different colleges and universities, those facts and principles that have a direct bearing upon the growth, learning and adjustment processes.

– Editor

CONTENTS

1

GROWTH AND DEVELOPMENT OF CHILD

Growth means increase in size, and skills. It also means improvement in existing abilities and skills. For example, a six month old baby has to grow in height and weight and has to sit up, stand, and walk gradually. Any new skill is clumsy and imperfect in the beginning. But with increasing age and practice the infant is able to perfect. A great deal of growth and development tin children occurs naturally. But there are several skills and activities that have to be taught. For example every child naturally learns to sit, crawl, stand, jump, walk and run, but only those learn to play tennis, or dance or swim who are taught these activities. Let us now see how children grow and develop during different stages of growth.

For nine months the baby stays inside her mother's body. During this time she grows and becomes ready for a life outside. Two little cells as tiny as a dot meet and multiply into innumerable cells. Some of these cells develop into different parts of the body i.e., head, body, arms, legs, stomach, heart, brain, bones, muscles etc. to finally take the shape and size of a baby. Two years after birth, the baby is called an infant and the period is known as infancy. During this period the infant changes remarkably from being someone who can only keep lying on her back and needing adults to attend to all her needs to a little child who is ready to explore the world outside the home. She increases in weight and height and learns a number of motor skills. She can hold her head erect, sit up, crawl, stand and walk.

She also knows how to use her hands to reach out to something, hold it and put in her mouth. She loves to play the game of dropping things if someone is there to pick them up. The skills of walking, balancing, climbing and use of hands in feeding oneself are achieved during the second year of life. By this time the infant also starts to speak a few words.

Meaning of Growth and Development

For some, growth denotes a progressive change in quantity, size or volume. The heart grows in size. It becomes bigger and bigger. The bones grow longer, thicker and heavier. In general, the body grows measuring more inches in height and more pounds in weight. These authors say that growth can be measured quantitatively in points, inches or pounds. *Development* also indicates increase but refers primarily to progressive changes resulting in improved quality, character and composition. For instance, bones grow in size, but they also develop when they change their composition and form. The brain cells grow larger but also develop by undergoing chemical changes. Thus development indicates more advancement, unfoldment, progressive changes both in quantity and quality and a going forward to greater maturity.

But this is only one point of view. Growth and Development are very closely related terms which, more often than not, overlap. But the whole fun is that each author wants they should mean exactly what he wants them to mean. To quote some of the statements that are noteworthy:

"Growth in an organism is that structural change which increases its level of functioning in specific ways," say Wilson. In so far as Jordan is concerned, "Growth on the otherhand will emphasize that progress toward maturity which is a result of inner maturity and environmental influences." And Woodworth and Marquis are of the opinion that "thus growth of any organism from an immature to a mature state is called maturation", and "development consists both growing and learning". "Development is a process resultant from a constant flux or interchange of energies within an organism and energies within its environment," says McGrow in the text edited by Skinner. Helen Thompson remarks: "Meredith concluded that the three terms, growth, development and maturation which are used in contradictory ways by different workers serve no useful and significant distinction' as their proponents had claimed. He proposed using the terms interchangeably, giving preference to growth as being shortest of the three terms."

From the discussions carried on so far, the present author concludes thus: Growth in human beings may be an inherited capacity and human development gets determined by a complex of inherited and environmental forces; but these two are closely related concepts—so closely related that though they may be distinguished in idea, they cannot be separated in actual performance. Each of these two shares in the process of helping an individual in his progress towards becoming the kind of adult that he eventually will be. What he is as an adult depends on how well he has grown and also on what he has learned. Therefore, the safest course to take is to use the terms interchangeably as Meredity suggests.

Childhood

During childhood i.e., from three years onwards, the child not only perfects the skills already learnt but also learns many more. She can now climb the stairs, almost like an adult, run, jump, eat, her food, speak to people in complete sentences, play with materials, build a tower of 10 blocks. With the addition of each year of life these motor abilities become more smooth and easy to accomplish. Can you think of the things an eight year old can do? Can she comb her hair, put on her clothes and button them? Can she play hopscotch, thread a needle and sew a button? Watch ten year old children fighting. They can box, punch and run away they can also carry things, even heavy ones, from one place to another. You realize that they have now become strong and capable. Besides the physical growth that is described here, a child also grows socially, emotionally and intellectually. By the time she is 3 years old she is enrolled in a school. Here she meets many others like her. She makes many friends and learns to play or work with them and share with them. Often there are fights and reconciliations among them but all this is a part of her social development. Emotionally, she grows by showing many more emotions than only pain and pleasure and also a greater control on these emotions. Remember, when she was an infant, she only cried or smiled? Now during childhood, she also shows fear, anger, jealousy, distress, love and affection, etc. As she grows in age her expression of emotions also matures. Can you recall the reaction of a two-year-old child when her attempts to achieve something are thwarted? How does an older react in the same situation? In school, a child also learns to read and write. She increases her knowledge and learns to solve various simple and complex problems of day-to-day life. She can gradually manage to handle some of her problems herself. This is her intellectual development.

There is one more fact that we think necessary to tell you, children show many individual differences in growth and development. You know that every child grows in height and weight but some grow earlier than the others. In other words every child shows changes in physical, social, intellectual and emotional aspects but the time of these changes differs. You can observe infants and children around you and see when each one starts sitting up, teething, crawling, standing, walking jumping, climbing stairs, etc. These differences, emerging in each infant and child at different stages of development are called individual differences.

A child is born and brought up in a family. What is a family? What does it do to a child? How does it help in bringing her up? These are some of those questions which will be answered in the chapter. A family consists of parents—mother and father, and the child, if she is the first born. If not, then there are one or more older children also. A family with only parents and their children is called a nuclear family. Quite often the father's parents, brothers and sisters also live in the same house. The brothers may be married so that their wives and children are also be staying in this house. A large family like this is referred to as the joint family. The system of living in a joint family is more common in a village than in a city. You may recall that when a child is born she is completely dependent on others for care and satisfaction of all her needs. The growth and development that takes place during early years and the care and protection provided by adults enable her to stand on her feet. The members of the family interact with her and over a period of time establish their own relationships with her. In the following pages your will learn about the role of these members in the upbringing of a child and the kind of relationship that develops between them.

Mother and father do so much for their children that is difficult even to list those acts. However, if you ever tried to list these activities you will realise that these can be of three types: (i) physical care and protection; (ii) satisfaction of emotional needs; and (iii) socialisation of children.

Care and Protection

Mother gives birth to a child she instinctively starts taking care of her. She protects her from external dangers, feeds her, and looks after her other physical needs. Often, the father also helps. In a nuclear family, where both the parents are going out to work the mother alone cannot manage to do everything. The father has to shoulder some of the responsibilities of looking after the children. Next time when you happen to visit such a family, do observe and note how the work is divided between the two parents. Also observe the problems when it is not divided.

Significant function served by the parents is that of providing a haven of love and trust to the child. From the very beginning the child needs the reassurance that she will be loved, cared and protected from threatening situations. The mother performs this role quite naturally as she brings her up. She conveys her fondness to the child during activities such as bathing, feeding and playing with the child. She constantly talks with her, sings to her ad pretends to get angry when the baby wets herself. The close physical touch whenever the child is picked up or fed helps to satisfy her need for love. When father takes over some or all of these tasks form the mother he is also fulfilling the same functions and establishing similar ties between himself and the child. When a child receives adequate love from her parents in her early years of life she is likely to develop into an emotionally stable adult.

Socialization of children means teaching the child the ways, beliefs, values and traditions of the family and society. In the process of bringing up their children parents teach them to acquire the ways, traditions, beliefs and values of the family and thus socialize them. They teach the children to respect all who are older to them, not to fight or be rude or aggressive, to be honest, take care of their things, to remove footwear when visiting a religious place or while attending a religious ceremony and to respect all religions and so on. What and how much children learn abut life also depends on the way they have been disciplined by their parents. Disciplining here means use of reward and punishment to correct their children's conduct. When they do not wish the child to repeat certain behaviour such as abusive language, they punish her. If on the other hand, desirable behaviour is shown, such as helping in the house, the child is rewarded through praise, privileges, affection and gifts etc. Through such interactions between parents and children, the children become acceptable members of their society. They learn their current ad future roles and the culture of the family is transmitted form one generation to the other.

Brothers and sisters, referred to as siblings, form an essential part of Brothers and sisters, referred to as siblings, form an essential part of the family. Their role in socialization of the child is secondary only to that of parents. In a family where there are two or more children and no other adult than the parents, the older child takes care of the younger ones in a way similar to that of the parents. Presence of siblings in a family is important for learning to share, trust, cooperate and love. The birth of a second child in the family is a painful experience for the first born because he has enjoyed undivided attention of his parents and others. But the first born soon learns that having a sibling, the same age mate, in the family is also a lot of fun. She soon finds that the love and attention from parents can be shared. Siblings provide one another emotional support, friendliness and companionship. They satisfy each other's needs for love, dependency and comfort. An only child often feels lonely without the agemates in his home and may even grow up to be a self-centered individual who finds it difficult to get along with others. When children grow up in a joint family there are not only more of the agemates for company and friendship but also more adults to depend on and to provide security. The joint family has a very important social role. Children learn to relate with their kith ad kin for whom there are different names in different Indian languages. Thus in the North the child learns to differentiate between his *bua* (father's sister), *Chachi* (father's brother's wife) and *Mami* (mother's brother's wife). She also identifies Mama (mother's brother) *Chacha* (father's brother). Growing up in close contact with the children of these uncles and aunts contributes towards social development of the child, especially if a child has no siblings of his own.

The thought of the grandmother fills one with memories of indulgent childhood and delicious sweets. In our country, the grandparents form a part of the child's environment from the time she is born. In fact, the postnatal care of the mother and the baby is frequently taken by one of the two grandmothers. From then onwards, the new mother learns many ways and beliefs regarding childcare from the child's grandmother, who is seen as worldly wise. Grandparents seem to enjoy participating in their grand children's upbringing. For one thing, they may have more time than the child's parents. Apart from that, with maturity they become more relaxed in life and hence are quite playful. In a joint family, a young child's care taking is jointly done by her parents and grandparents. The mother, being occupied with household work, hands over the baby to one of the grandparents. The child spends several hours in a day with him or her and this develops into a bond between the two. Hence it does not mean that children growing up in family patterns and role descriptions different from those described above are in any way at a disadvantage. What can be stated here definitely is that all children grow up in families.

Need for Child Centred Education

Today a school is what you know. The government of our country under the constitutional directive, is obliged to provide free and compulsory education to all up to the age of 14 years irrespective of their caste and creed. The aim of these schools is to provide opportunities to each child for learning of three 'R's and also for an all round development of her personality. Each child must learn to be responsible, cooperative considerate, tolerant, patient and independent.

In school when she in tracts with others of her age or older and younger children she learns all these traits. Children who come to a school are not alike. Some are bright and gifted, some are inclined to be slow or take their time to learn. Some students are creative and like to work in their own style. Schools should also be able to cater to individual needs. Education does not mean fitting in to a box or learn to be like everyone else. Learning in a group need not put an end to individuality.

(i) Authoritarian: Students have to do what they are told. They do not have any right to question.

(ii) Authoritative: There is guidance but students have the right to make choices, ask questions.

(iii) Permissive: Students can do what they want. You might feel the third to be the best. But think of it, if everyday you were left to do what you want, would you be able to cope with such a situation? Science, History, English, names of books can you think of it all? It has been observed that students do like some amount of guidance, a framework within which they enjoy freedom and flexibility. In some of our schools we always hear, 'Be quiet. I am telling you'; 'No talking', or 'put your finger on your lips'; etc. a five year old boy attending a highly regimented authoritarian K.G. told his mother 'Teacher says to put your finger on your lips, I do that but still I can talk'. This is an instance where too much rigidity leads to teaching children to cheat or learn manipulative tricks. Of course children cannot learn admits noise but children also cannot be quite for long. So what is the solution? Teaching processes should be interesting and children should feel involved an not be idle or distracted.

In the early years it is extremely important for learning to be meaningful. We cannot teach our children I for Ibis and

P for Penguin. These are birds our children never see. Preschools and nursery can not be too structured either. Nursery school is the first exposure to the outside world. It is important that children enjoy these early learning experiences. This is the reason you see colourful pictures, posters, toys, swings in these schools. They have attractive names like 'Happy Hours', 'Tiny tots', 'Stepping Stones', 'Little gem' etc some of these schools focus on social development. Some accelerate cognitive growth. Some schools have a daily structured programme whereas some let the children decide the activity for the day. It is recognized that preschools have to be different from classrooms in later school years. There is a caution that too much freedom tends to cause as much harm and confusion as too much rigidity can lead to frustration. Learning occurs best when the child a part of a group of discoverers, plays, questions and gets guidance. It is important to provide experience and give meaning to it. The teacher is a leader who plans, provides and guides.

Teacher's Role

Most of us cried and took time to adjust. For the child form the very beginning it is important as to who she meets in school. Teacher is a substitute mother. Many young children make mistakes and feel embarrassed that they called teacher 'mummy'. It is an error but one that has deep meaning. There was a 4 yea-old child who even after three months of attending school cried often, hesitating to go to school. She always said, 'teacher never smiles, I want the teacher to smile.

It is important for the teacher to e attractive and charming. It has become almost customary for teacher to talk in a high pitched voice forgetting that raising your voice is only a short cut to achieving respect and discipline. A gentle smile and a hug can get much more done. Needless to say teachers need excessive patience. One way to reduce the work load is by not having more than 25 children in one group. In small classes, children get attention. A touch, a pat on the cheek can do wonders in student's behavior. How well students do greatly depends on what the teacher thinks of them. Students can feel what the teacher expects and how she feels towards them.

Gradualy, the child's ability to make different sounds increases. The development of the ability to think in a child takes place instages in the same way as it happens with other areas of development.

2

PRINCIPLES OF GROWTH AND DEVELOPMENT

The human baby of today is neither the same as that of yesterday nor the same as that of tomorrow. It is capable of growth. Though slowly yet constantly it grows. Human Development is Determined by a Complex of Inherited and Environmental Forces. In the fourth chapter of this text it was explained how an individual at any stage of his development is the product of hereditary and environmental factors working simultaneously upon him. Also it was explained how an individual reacts and behaves in all life situations in response to those two interacting forces.

The child behaves as he does because he is a human child with an array of needs demanding satisfaction. He is surrounded by socio-cultural forces which determine how these needs should be satisfied. The endocrine glands shape his mental and emotional life primarily, and so on. Therefore, this principle of growth and development does not need an elaborate discussion here. We shall be returning again to this principle in later chapters.

People generally divide the process of growth into different stages such as : Infancy, Childhood, Pre-adolescence, Adolescence and Adulthood. This does not mean that there is a gap in the process of growth between one stage and another. Growth does not leap from one stage to another. It is not that a boy grows three feet in height, then his growth stops for sometime and then suddenly the boy begins to grow again in height and reaches a height of four feet six inches. Growth is gradual and continuous. At each moment of life any individual is in the process of changing into something a little different from what he now is. Life begins as a single celled organism but by cell division grows into millions of cells; and the organisation of bodily cells become more and more complicated. The mental and the speech abilities of the human child grow slowly but continuously. Everyday the child acquires knowledge of different kinds in different degrees. At first, the baby babbles and gurgles, but gradually he begins to talk and then read. As the human child grows, his behaviour pattern grows more and more complex in nature. When the baby is hungry and milk is not given to it, naturally, it would cry and roll on the floor. But an adolescent even through food is not given to him when he is hungry, shall not cry and roll on the ground. He shall express his resentment in a different way. This is because the adolescent has made an advancement over the baby, both emotionally and socially.

The Rate of Growth is not Even

Of course, growth is gradual and continuous; but, it is more rapid at some periods than at others. In infancy the rate of general physical growth is more rapid than at any other stage of development. The organs of the body mature at different rates. Legs grow faster than arms. Brain grows faster than sex organs. The sex glands grow at a different rate from the adrenals or the thymus. Hence one individual differs from another on account of differing stages of development reached at by the bodily organs.

Different traits of an individual develop at different rates and reach the limits of development at different times. In the same individual various components of personality may grow at different rates. These differing patterns of growth in the same individual are referred to as "*ages*". The child has many ages. He may be ten years old chronologically, have a mental age of 12, and a social age of 14. He may have an emotional age of 13 and an educational age of 11. The rate of maturation differs from individual to individual.

Maturation refers to the inner changes that occur in the individuals as a result of heredity and irrespective of environmental conditions. In whatever environmental conditions they may be, at first the babies are not able to sit but a time comes when they sit up without assistance. And another time comes when they begin to walk and talk. It is on account of the process of maturation that such things happen. Differing rates of maturation and unevenness of growth are of great educational significance as they account for individual differences.

In our country, our greatest drawback in the field of education is that we overlook individual differences in growth and development. We admit pupils to different classes on the basis of their chronological age and ignore their respective levels of development, mental, emotional, social and physical. While admitting him into a class we do not consider the different "ages" of the child. We do not group children for instruction according to their level of mental development. As a result, the children face a lot of difficult in adjusting themselves to school instruction and environment. The bright boy doesn't like the mode of our instruction and the dull boy is unable to follow what we teach. Each child grows at his own unique rate. Therefore, uniform treatment of children at any level or in any subject is quite unsound. Children respond differently to the same learning activity or the same subject matter or the same teaching procedure.

Therefore, individualised instruction is a necessity.

Curriculum and teaching procedures must suit individual children. Only an efficient and skillful teacher can do this. One should not expect the same level of achievement at the end of a teaching-learning course. The achievement of each child shall depend upon the forces working within him and the forces acting upon him from without. Also the nature of personal-social-emotional adjustment of children vary both in kind and degree. Therefore, to treat all the children alike in a teaching-learning situation is educationally unsound and unwholesome. Some suggest the classification and grouping of children on the basis of their IQs. Others do not prefer the classification of children on the basis of their mental age. They prefer *"individualised instruction"* to "ability grouping." The reason for this is that children grouped under the basis of mental age may widely differ in their emotional age or reading age or social age. If the teacher wants "ability grouping," he has to classify children in one way for one purpose and in other ways for other purposes. Therefore, "individualised instruction" is the safest course they argue.

Growth is Rapid in the Early Years of Life

The rate of general physical growth is more rapid in infancy than at any subsequent stage of development. During the first six months of life the rate of growth is most rapid. In the early years of life the growth of body makes a rapid progress, then the rate of growth slowers down and again when the individual is about 14 years old, the body begins to grow very rapidly. But the time the child is about 2 to 3 years old, his sense organs function very well. When he is six years old, his nervous system is nine-tenths of the adult weight at birth its weight being one-fourth of the adult weight. As studies by Thurstone and Thorudike show, mental development is very rapid during the early years of life.

Rapid development during the early years of life has some educational significance. As a number of functions develop together in the early years of life, the young child wants to do a number of things he is capable of doing. Therefore, the school should engage children in a number of activities that the children are ready to perform. In order to encourage children to display all the capacities they have developed, the school should provide them with a rich environment. While the children are young, they should be helped to have better personal and social adjustment, through a variety of activities they shall be engaged in.

General Pattern of Development

Each one of us is an organism. We belong to the human species. Qualities and characteristics grow and develop in us in orderly succession. The orderly succession of the growth of traits in us does not differ from individual to individual. The infant lifts its head first, then its chest and then its stomach. The baby sits, crawls, stands alone and then walks in that order. When he is five years old, a boy can copy a square and reproduce a diamond when he is seven. If he cannot do the former at five, he cannot do the latter at seven. When a small cube is presented to infants who are 16 weeks old, they shall only gaze at the object. But if the cube is presented to them when they are 40 weeks old, 85 percent of them shall surely try to touch it. The sequence of development is orderly and predictable. If you find a particular trait has developed in a child, you can tell what shape it shall take in future.

In the early stages of human growth, the direction is from general to specific or from mass to individualism. In the beginning, there is a total movement of the body. Gradually specific actions follow and the total movement stops. After they are born, infants for about 10 to 15 days move their whole body to stretch or sneeze. After sometime, when infants try to lift their heads, they exercise all the muscles of the body. But when they begin to walk after a few months, they use only a few muscles. They walk with the legs and use hands for other activities.

The behaviour of the neonate is a mass activity in which the whole body participates; but gradually, specific parts of the body are used for specific purposes. At first, the responses of the baby are generalised. In course of time generalised responses give place to specific responses. For instance, the cry of the baby in the beginning is a generalised response. His cry becomes a specific response when he cries only when he is hungry. The baby uses the word "Daddy" to greet most of the male members of the family. This is a generalised response. It becomes specific when the baby uses the word Daddy" to greet his father only.

Differentiation and Integration

The direction of growth from general to specific or from mass to individuation is the process of "*differentiation*" But development takes place through the process of "*integration*" also. "Many activities as children grow older are integrations of specific processes already acquired" says Jordan. "Walking' is an excellent example to show that development is a process of integration. Simpler habits of the arm control, leg control, control of body movements and balancing are combined into the integrated activity of walking. In games of skill, there is an integration of many learned habits.

For example, in a game of football, quick starting, swift running and stopping, easy control of the round object by the right foot or the left and many other specific processes are integrated and arranged into one organised whole, a complex pattern of skill.

Growth and Organism

Growth relates to the whole organism—the total personality. It is the total child who grows. Physical, emotional, mental, social and moral development occur simultaneously and interact upon one another. They are interrelated and interdependent. However, physical growth is fundamental forming the base. Along with the body the nervous system grows. Mental development depends upon the growth of the brain. Language development is dependant upon the growth of the brain, eyes and ears and the vocal organs. Language development shall get hampered if the brain and the sense organs have not developed well and do not function properly. If his language development gets hampered, it would cause emotional disturbances in a child. Children with physical defects and deficiencies are emotionally disturbed and find it difficult for personal -social adjustment. Even they may not get accepted by the same age group.

Growth is of the total individual and the function of education is to elevate the entire personality of the child. Teachers should provide the most favourable environment to stimulate the growth of the whole child. They should engage children in varied and novel activities to draw out the inherent potentialities of the latter.

(i) Design of Development is Individualistic

Individuals differ at birth. Each child possesses his own growth and development pattern which tends to remain constant from infancy to maturity. Children who are tall or short in the early years of life continue to remain as such till adulthood. The dull or bright continue to be dull or bright. Each child grows at his own unique rate and even though environmental conditions change, the rate of growth remains relatively constant. This trend in growth and development pattern is of great educational significance.

It is possible to predict the extent to which mental development shall take place in a child as he grows older and older. Also it is possible to tell in advance the kind and degree of education the child shall fit into. A study of the child's present behaviour will tell us the nature of his behaviour in future and what course it is likely to take. This principle, if they are aware of it, helps teachers to shape the behaviour of children.

Behavioural Changes Result from Learning

Maturation stands for the genetically determined growth. It refers to that phase of growth which relates to the unfolding of whatever characteristics are incorporated in the genes. Babies sitting up, crawling, walking and the bladder and bowel control are products mainly of maturation. Changes in behaviour result from learning and maturation, both in important amounts.

"Learning is a process or operation inferred from relatively permanent changes in behaviour that result from practice. On the other hand, relatively permanent changes in behaviour that result from maturation and temporary changes that result from drugs, fatigue, and the like are not considered to be learning" All the children, stand upright, walk or run rather than walk when the time comes, without practice or instruction. Such things as these result on account of maturation rather than learning. But take into consideration a young boy learning the skill of riding a bicycle or a young lady learning the skill of walking on a rope in circus. The boy and the lady mastered the skills as a result of learning. Learning is the result of reinforced practice. A 6-year old child cannot recognise any of the twenty words in his First Reader. He receives instruction and practices. Six months later he recognises all the words. He could recognise all the words as a result of reinforced practice. Therefore, reinforced practice is the cause of learning and learning is the result.

However, maturation and learning are so closely related that their influence cannot be isolated. They operate as dual forces in all cases of behavioural change. Children grow into learning. In the case of an immature, still-growing individual, it is too difficult to make a distinction between maturation and learning. As a result of reinforced practice, the young boy is able to learn the skill of riding a bicycle; but he cannot learn this skill until his bodily organs have reached the required level of maturity for learning this skill. Therefore, the developing child learns while he grows and grows while he learns. In the early stages of development learning and maturation go hand in hand.

Maturation provides the necessary physiological foundation for learning. It is considered as an essential prerequisite for learning. Performance of different school tasks require different levels of maturation in children. But it is not possible to hasten the process of maturation and make the children ready to learn certain school tasks that the teacher desires they should learn. However, reinforced practice has its own value in the early years of development. In the first place, such practices result in increased self–confidence and the second place, children feel proud when they perform a task successfully after considerable practice. This makes them learn desirable motives and interests.

The teachers should be cautious in so far as these issues are considered. Certainly, a considerable portion of a child's readiness for learning depends upon maturation. On the other hand, environment also plays its own part. Earlier it was discussed how an unfavourable environment causes permanent set back to maturing intellectual functions and a rich environment boosts upmental maturation during early years of development. The teacher need not always wait the child to be ready for learning in the normal course. He can motivate the child to learn certain tasks providing a rich and favourable environment. Practice in the development of certain skills during the early ages is a necessity. For example, language skills and particularly learning a foreign language, can be profitably accomplished at a much earlier age than at the high school level. Practice of reading skill is necessary for the normal child during the early years of development. Certain mathematical concepts can be developed much earlier than was once thought possible.

3

CHILD BEHAVIOUR AND ENVIRONMENTAL INFLUENCES

The word environment means surroundings. For every child there is an environment in which she grows up. Her surroundings include the family, neighborhood, society, culture, religion and also where she lives. Part of the process of growing up in having experiences. Environment provides opportunities for a child and these influence her growth. Every child lives in a home, talks to people, goes to places, watches other children play, celebrates festivals, goes to the market and does many more things. The more varied her surrounds and access to them and freedom to interact with them, the richer is her experience. Let us now examine the contents of the environment one by one.

At home there is a family consisting of parents and siblings and/or of grandparents, uncles, aunts and cousins. When they interact with the child it adds to her experiences. The interaction is different when the family is larger than when it is small. Are you living in a nuclear family? Imagine yourself in a joint family and compare your reactions and interactions with each member of the family. Suppose there was a pet too in the house—a dog or a cat or a parrot. What would have been your interaction with it? Besides people, a home also contains a number of material objects. Some of those are absolutely necessary. For example, a stove and a few utensils used for cooking, a bed and bedding to sleep and protect oneself from cold, clothes to wear, etc. If you made a list of these objects in your homes and compare it with items in some other homes you will realize that the number of items vary from house to house. Some have very few and others have many more items. Most of these objects are there for comfort. Among these are the children's toys with which they play and spend their leisure time. Some children also adapt objects to become toys: a stick can become a horse or a plastic container becomes a train. Do you realize that the experiences of a child are richer if there is a variety in the objects a child is exposed to? There would be differences in the experiences of a child growing up in house in a slum area and another in a house with plenty.

School who are the people and what are the material objects. A Child who is enrolled here comes in contact with all these. She meets teachers, peers and other adults. She develops her social skills and learns to share, cooperate and be part of a group. By interacting with equipments and materials in school she is able to strengthen her physical and motor co-ordination. This means that if the school has a good programme, facilities to interact with people and exposure to a variety of things, the child has plenty of opportunities to develop in all aspects of her personality—physical, social, emotional and intellectual.

One of the major aspects of schooling is to provide an opportunity for children to be with their own age-mates. Children enjoy this and they always seek company. When they are together they might fight over toys, sharing food but still they like to be with each other. Company of children is important in the life of the child. Peers provide a more realistic model for developing skills and attitudes. A child learns to climb a tree, by watching an adult but gains more confidence when he sees his friend do the same. Peer group helps children form attitudes and values. Everyone at school talks about a T.V. programme so it must be watched. Children do what their friends do. This gives them a feeling of belonging to a group. Peer group also provides a medium to test the values you derive from your parents. As a girl you might not be allowed to go out alone after dark, you get satisfaction that your friend has the same rules to follow. Peers offer emotional security in terms of similarity in thoughts, problems and ideas. You do not like mother when she scolds or when she forces you to drink milk. You feel relieved and less guilty for pestering you mother when your friend has had similar experiences. Sometimes behaviour of children is moulded to suit group behaviour. There was a five old child who spoke very good English. But soon her diction changed and style become influenced. By Indian dialects because everyone at school spoke like that. When families go to a new place with different language, it is the child who learns the language first. Children learn games, songs, folk lores, jokes from each other. This learning comes only from friends.

Culture of Childhood

Games like marbles, hide and seek, kho kho, seven stones, skipping rope, hops skotch have existed for a long time. They may have slight alterations but basically each generation passes it to the next,

Rhymes no school teaches but children learn. They could be a parody of adult songs too. Rhymes provide an ease for frustration and aggression or merely letting your anger out in a manner that is easy and not unpleasant. Of course this happens unconsciously. They also highlight the dominant cultural values and significance of adult family members.

Behaviour refers to *observable activities* and *conscious experiences* of an individual. There are certain activities of an individual that can be observed by another individual. For example, I can watch you reading or riding, eating or playing. Some other activities can be observed by an experimenter's instrument. A doctor can observe the activities of your heart by means of an electrocardiograph. You are angry at something or Somebody. You alone know what it feels like to be angry. Or, you praise your student for some good work. He alone knows what it feels like to be praised. An individual has his own *Conscious experiences*. He has his own joys and sorrows, private thoughts and dreams, of which he himself is aware. These are Conscious mental processes. The individual can make a verbal report of his conscious experiences. Thus, conscious mental processes are accessible to others through verbal report or by way of inference from other behaviour. Any form of reporting on subjective conscious events or experiences is *introspection*. Included within behaviour are verbal reports made about subjective conscious experiences.

A process is a series of activities. Besides conscious mental processes, *unconscious mental processes* exist in us. Generally, we are not aware of them. They are not readily accessible to voluntary control. They may prove harmful if we become aware of these unconscious mental activities. The contents of the unconscious come from two sources. A part of it are the individuals inheritance. They are primitive brutal ideas and strivings that have never been conscious. The second part consists of buried thoughts, emotions and impulses, memories and wishes that were once conscious but have been pushed back into the hinterland of the mind because they were too shocking, painful or shameful to tolerate. While the conscious mental activities are internally consistent, the unconscious mental activities are internally chaotic.

From the discussion carried on so far, all the activities of an individual are classified as observable activities, conscious mental processes and unconscious mental processes. Thus, psychology is the systematic study of the fundamental laws that govern the observable activities and the conscious and unconscious mental processes of an individual.

To serve our purpose, we may take into account two definitions of Psychology. One: "Psychology can be defined as the science of the activities of the individual." Two: "For our purpose we will define psychology as the science that studies behaviour and mental processes." Both these definitions mean the same thing.

What is Education?

To define education is not that easy. Complexity of human nature and environment, different schools of psychology and philosophy, different political ideologies and cultures lead to different definitions of education, ultimately turning it into a rainbow. The existing definitions we come across are too broad or too narrow; each definition is tinged by a particular 'ism' or it emphasizes a particular aspect. While some believe education to be a life-long process, others limit it to instruction and training in the academic institutions. When Gandhiji speaks of education as, "By education, is mean an all-round drawing out of the best in the child and man—body, mind and spirits," he emphasizes upon the inner potentialities of the individual. On the other hand, when Dewey say, "Education is the development of all those capacities in the individual which will enable him to control his environment and fulfil his responsibilities," he emphasizes upon the social and environmental aspects of education. What we need is a sharp and precise definition of education, useful and if possible, functional in nature that should fit in the theoretical framework of the present text.

Of course, Peters impresses upon us the difficulties involved in aiming at such a definition of education when he says: "in exploring the concept of education a territory is being entered where there are few signposts. To use Ryle's phrase, the 'logical geography' of concepts in the area of education has not yet been mapped...'education' is a concept which is not very close to the ground. Yet, we have to choose a definition to suit our purpose.

So, *'education' is a "Concept"*. "Equality," "justice," "democracy" "education"—all these are concepts. When a symbol stands for a class of objects, events or ideas with common properties or relationships, we say that it refers to a *Concept*. "Vegetable" is a concept based on common elements. "Equality" is a concept based on common relations. Concepts may be of concrete things, for example, the concept "stallion" refers to a given variety of horse. They may be of abstract ideas, for example, "equality, "justice," "mass", implying relationships common to many different kinds of objects or ideas.

To express the same things in a different way, A Concept expresses an abstraction formed by generalisation from particulars. "Weight" is a concept; it expresses numerous observations of things that are more or less "heavy" or "light'. A Concept that interests us much is "achievement," it is an abstraction formed from the observation of certain behaviours of children. These behaviours are associated with the mastery or "learning' of different school tasks—reading words, doing arithmetic problems and drawing pictures. The various observed behaviours are put together and expressed in a word - "achievement." A definition consists of a statement of equivalence between the concept to be defined and a set of other concepts whose meanings is presumed to be known on other grounds. This means, we can define one concept with other concepts.

For instance, we can define "weight" by saying that it is the "heaviness" of objects. Or we can define "anxiety" as "subjectified fear." In both cases we have substituted one concept for another. Phenix develops a definition of education using six other concepts in the following manner. First, education is a *"human enterprise."* It is confined to human beings only and has nothing to do with inanimate objects, plants or animals. Man is distinguishable from all other things, both living and non-living on account of his special nature. We can train a dog but we cannot educate it. We can surely educate a human being.

Second, education is a *"process."* It is not a thing which persons possess or which is a part of them. You can possess a pen or have black hair, but not education. Even knowledge, skill or character are not in themselves education. They stand as evidence that one has been educated. Education is not a thing but a process. It is a kind of activity or a series of activities in relation to human beings. To educate a boy or girl is to engage him or her in a process. If you say, you 'have an education', it means that you have undergone a process. Thus, education is a *"person-related process."*

What kind of person-related process education is? Third, education is concerned with the *"development"* of human beings as persons—physical, mental, emotional and social development. It is a process involving personal becoming, and growth resulting from personal experience. It is concerned with the real an in- most nature of the person. The process of acquiring property or becoming famous is not education. Education is a process of personal growth or learning and also a process of bringing about changes in people for good.

Not all processes involving human development are education. Genetically determined growth, that is, maturation is not education. A need is a state of tension forcing a person to activity; but it is not education. Education is only those processes which *guide, direct* or *fashion* personal becoming. Education requires an educative agent acting upon the person to be educated. Therefore, fourthly, education is an *"interactive, bipolar process."* From this it is obvious that this definition of education does not include in it a view of education as a process of unfolding the innate potentialities of a person already present but not apparent.

Pets do help to fashion children's growth as persons. The physical environment of human beings also guides their learning. But the proposed definition of education does not include these as educational agencies. According to it education is restricted to the *direction, by persons,* of personal development. That is to say, the only admissible agencies for education are human beings. Included in such personal agencies is the self. The individual can direct his personal development. This is called *self-education.* It was sated that education is an interactive bipolar process. It is the nature of the self to be bipolar. It has the ability to carry on an internal dialogue. The philosopher or the scientist often does converse with himself. The dialogue may be within one person and if it is both solitary and silent, it becomes *reflection.* All reflective thoughts are inner dialogues. As such, the fifth qualification of the proposed definition is that education is a *"process by persons."*

Finally, the qualification that completes the definition is that the direction which persons give is *"intentional"* or *'deliberate."* Friends influence us in many ways; but such influences do not come under education. If a friend deliberately seeks to direct the development of his companion as a person, he becomes an educator. Not all intentional directing of persons can be education. The traffic policeman intentionally directing the automobile driver to slow the speed or take a turn or stop is not education. Such direction does not bring about the growth of the driver as a person. The intent is to get the traffic move smoothly. Only those deliberate processes of directing which are intended to influence personal growth and learning fall within the definition of education.

Thus, the definition of education which has been developed step by step is: *"Education is the process whereby persons intentionally guide the development of persons."*

This definition of education fits in well with the theoretical framework of the present Text; and hence it is accepted without any reservation.

Educational Psychology

"To state the two definitions again with brevity:

Psychology is the science of human behaviour and mental processes; and Education is a person-related developmental process, interactive, directional and deliberate. From these two definitions, can we not deduct the *nature and functions* of Educational Psychology?"

A student jumped upon his feet and asked: "But how?" The teacher said: "I withhold the answer for the present." At birth the child brings his biological inheritance with him into the world and comes in contact with an environment. The environment is saturated with social culture. The characteristics of the biological heredity of the child, in themselves are not sufficient to enable him to develop harmoniously in the social culture around him. He has to be helped in acquiring the skills, information, concepts and attitudes concerning his social heritage. Therefore, society provides a variety of educational institutions to stimulate and nourish his development as he grows and learns. The child has to be socialised.

Education is a process of socialisation through which he has to be socialised. "A half century of research has given us a very different picture of the nature of the child from that held in earlier periods. The following brief summary....outlines the essentials of the modern point of view. (1) The child is a growing organism who moves by infinite small changes from a single cell to a complex organisation of structure and function at an adult level....(2) The child is a spatially separated unit that functions as an organised whole in the situations he meets...(3) The child lives in a context, itself neither simple nor unitary, which continuously effects his behaviour and development...(4) The child is engaged in an ongoing process of development that is not reversible." Throughout the history of education, scholars have set forth challenging theories and view points on the problems of growth and learning of the child whose nature is stated above from the modern point of view.

The beginnings of educational psychology are to be found within these. A lot of the theoretical frame-work that educational psychology incorporates in it, perhaps originated in the classroom where the teachers were confronted with problems relating to the pupils. Educational psychology is an applied psychology. It is especially concerned with the application of methods and principles of psychology to human behaviour in educational situations. "Although educational psychology is a relatively new science, its aims and content have become fairly defined. The aims set forth by a committee into the late 40's are still appropriate today."

"Educational Psychology is concerned primarily with the study of human behaviour as it is changed or directed under the social process of education, and secondarily with those studies of processes that contribute to an increased understanding of how behaviour is changed and directed through education."

Educational psychology is the systematic study of educational growth or development. In the study of educational psychology we try to ascertain the nature of educational growth, to learn its typical features, and the forces which affect it. Educational psychology is an area of knowledge, an application of methods and principles of psychology, and a field for continued study and research. General psychology and educational psychology are dissimilar in that educational psychology selects and employs certain data from the general field. The major contribution of psychology to education lies in providing a scientific foundation for the art of teaching. The primary aim of educational psychology is to contribute to an understanding of sound educational practices. Its primary concern is to promote the development of professional insights into the principles underlying the teaching art.

"Educational psychology provides the psychological foundations for education in somewhat the same way that physics provides the foundations for structural engineering ....educational psychology contributes to the foundations of education by providing an approach to educational problems, as well as a set of techniques for studying children and the problems that arise in their education. It similarly offers information concerning the developmental, learning and adjustmental processes of children. Educational psychology also provides the skills and techniques required to acquire additional original information, as well as to evaluate and interpret the research findings of others. It consists primarily of a point of view, a body of information, and a group of techniques and procedures."

In the passage quoted above, the authors emphasize on an important point, that is educational psychology is chiefly concerned with the development, learning and adjustment of children. Blair, et al., do the same thing when they say: "The teacher who would successfully guide the development, learning and adjustment of children must

(a) possess a comprehensive and integrated set of psychological principles which explain human behaviour,

(b) possess a technique for studying the individual child in order to determine which principles explain his behaviour in a given situation, and

(c) be able to analyse his own teaching and learning procedures."

The most acceptable statement of the relationship of psychology to the field of education is that psychology makes it clear that rules don't work in educational practices; what is right in one situation is, as often as not, wrong in the next. And educational psychology is oriented towards the application of the principles and techniques of psychology to the solution of the problems of the classroom. "Kelly has analysed the task of educational psychology into nine different purposes which may be condensed as follows:

(a) to give a knowledge of the nature of the child;

(b) to give an understanding of the nature, aims and purposes of education;

(c) to give an understanding of the scientific methods and procedures which have been used in arriving at the facts and principles of educational psychology;

(d) to present the principles and techniques of learning and of teaching;

(e) to give training in the methods of measuring abilities and achievement in school subjects;

(f) to give a knowledge of the growth and development of children;

(g) to assist in the better adjustment of children and to help to prevent their maladjustment;

(h) to study the educational significance and the control of the emotions; and

(i) to give an understanding of the principles and techniques of character training".

To summarise the discussions carried on so far in the words of G. Lester Anderson:

(i) Educational Psychology is focused upon the human behaviour.

(ii) It is a body of facts or information that has resulted from observation and investigation.

(iii) This body of knowledge can be summarised or generalised into principles or theories.

(iv) Educational psychology has developed a methodology by which investigations are made, information is discovered, hypotheses are tested and theories are derived.

(v) This methodology is useful in arriving at the solution of educational problems as they present themselves. (6) This information, this knowledge, these principles and this methodology constituting the substance of educational psychology, provide a base for educational theory and educational practice."

Educational Psychology draws material from different other branches of psychology, namely, general psychology, experimental psychology, child psychology, clinical psychology, developmental psychology, the psychology of motivation, learning and personality, abnormal psychology, psychological statistics, social psychology and the like, and employs them in solving educational problems. It helps the teachers make wise decisions in their search for solutions to different problems both inside and outside the school. Wide scale research is conducted in the teaching of different school subjects, measurement and evaluation of pupil qualities and characteristics, special education for exceptional children, guidance practices, remedial programmes and social psychology of the classroom. Educational psychology has contributed as much to general psychology as it has had occasion to borrow.

Educational psychology as an applied science is not directly concerned with educational goals. However, in atleast three ways it helps the teacher and school better handle the objectives provided by educational philosophy and society.

1) It helps define these objectives operationally, that is, to translate them into terms of implied action and measurement.

2) It provides experimental methods and evidence yielded thereform as to how objectives may be achieved.

3) It gives various means whereby the attainment of aims can be measured.

"The role of educational psychology in education may be summarised as follows:

(a) to evaluate educational theories;

(b) to study and evaluate contemporary practices in education;

(c) to evaluate different methods of teaching in the light of established principles of learning and motivation;

(d) to help teachers solve their personal problems; and

(e) to furnish methods to help those engaged in the scientific study of problems in the general area of education psychology."

From the discussions carried on so far, it is obvious that the teacher who has a good grounding in the field of educational psychology finds a rationale for the problems he faces and for the decisions he must make.

How does Educational Psychology Help Teachers?

All other educational problems revolve around the main three—development, learning and adjustment of children. In order to guide these three successfully, the teacher must know children individually, know a set of well formulated psychological principles and know where and how to use these principles. The teacher's position is similar to that of the medical practitioner who must first study his patient before prescribing for him; and a correct diagnosis is needed for a correct prescription.

The most important single factor underlying the success of a teacher is his personality and ability to relate it to the class. The best reminder he should take with him into his classroom is that learning is effective to the extent that it involves the goals and purposes of individual children. The teacher's major contribution to the maximum self-realisation of the child is best effected through sensitivity to pupils' needs, goals and purposes. Educational psychology assists the teacher to develop competence in studying children, in utilizing psychological principles, and in evaluating his own teaching methods. Within the school, there are three main tasks of the teacher to perform.

Selecting and Organising Material

Before he enters into the class to teach, or outside the class to help the children under his care, the teacher first has to *select and organise the material to be learned*, "Material to be learned" may include learning a date in history, reading a lesson in language, map reading in geography, typing or riding a bicycle, even learning attitudes, interests or a system of values. Certain material or experience will be presented by the teacher to the pupil at a particular time in the pupils' school life and in a certain order. This is not a simple task. The teacher has to ask hundreds of questions and answer them himself, and make wise decisions, before he takes any step in this direction. An English master asks the question: What English should I present to the beginners? How must grammar can they learn? A Mathematics teacher asks the question: Can a student of Class X master set algebra? A Science master asks the question : Is it wise to present the principle of Archimedes to children of class IV? A Geography teacher asks the question: Should I teach physical geography before I go to regional geography or vice versa? This process of selecting and organising material we call *"building the curriculum."*

If they are to do their work to the satisfaction of all and build up the curriculum, teachers must ask hundreds of such questions and find out answers for them. The answers should be based on specific information. Educational psychologists have gathered at least a part of such specific information and set down as general principles of child growth and learning. To accomplish his task successfully, the teacher should be aware of this body of knowledge and utilize it judiciously, as it becomes an indispensable part of his professional equipment. Without this body of knowledge educational psychology provides, he would fail in his professional judgements essential in selecting and organising the material that pupils are to learn.

Guiding and Directing

Education is possible because humans can learn. While defining "Education" it was stated that education is a directed and deliberate process. The teacher plans and guides or directs the learning experiences of his pupils in harmony with stated educational objectives. An '*Objective*' is a stated desirable outcome of education. The teacher guides or directs the learning of his pupils. This is a difficult task he has to perform. He should be conscious of the fact that children "learn" while they "grow", and grow while they learn. He should know how children learn and decide before-hand the activities he is to engage his pupils in to bring about behavioural changes in the desired direction. He should have good knowledge of the techniques of motivation of learning in general and classroom motivation in particular. He must be aware of individual differences in learning ability, judicious use of reward and punishment to promote learning and the extent to which learning in one situation gets carried over to another situation.

The major concern of educational psychology is learning. Most of the information available concerning the problems of how children learn, the things that motivate learning, and the many factors that facilitate or retard learning are of use to the teacher. Educational psychologists have formulated principles about the use of rewards and punishment. They can tell us something about the teaching aids we should use while presenting different lessons on different subjects to the children and what can be done to minimise forgetting. All these items of information and more are relevant to the teacher's task as one who guides the learning of others.

Evaluation of Pupil Qualities and Characteristics

In addition to finding out how far the educational objectives provided by the Nation, the Society and the School have been realised, the teacher should know how well he has performed his task as a teacher. Also he wants to know how well his pupils have learned the material he has presented to them. The pupils want to know their progress; and the parents want to know how well their sons and daughters are doing at school. The college the pupils desire to enter in want to know what they have learned up to date. The employers want to know how well the candidate has done in the school, if the pupil seeks a job. Ultimately, the public demand to be informed about the qualitative aspect of their schools. Therefore, the teacher must test his pupils and evaluate not only their achievement but other qualities and characteristics.

"*Evaluation*" means, first measuring a particular quality or characteristic of the pupils and next attaching a value judgement to the measurement. For measuring the achievement of his pupils or some other quality, either the teacher has to design and construct his own test or select and use tests designed and constructed by others. If he desires to use tests made by others, he should know where they are available. He should know how to administer the test, score it and interpret the results. Also he should have some knowledge of the validity ad reliability of the test. Validity refers to the degree to which a test is capable of achieving certain aims; and Reliability can be defined as the degree of consistency between two measures of the same thing. He has to make reports to students and parents. He has to prepare records which help to decide who should be admitted to college and who should not.

The educational psychologists have investigated these matters. They have developed techniques for increasing the accuracy of general observation of children. They have studied how to measure the qualities and characteristics of people and have made available for use a wide array of formal and informal tests of aptitudes, achievement attitudes, interests, personality and adjustment. They have devised techniques for recording information gathered and for interpreting it. They also use scientific methodology in increasing knowledge in educationally significant areas. The teacher must study the psychologists, findings to equip himself to perform this task of evaluation with professional skill.

Personal and Social Adjustment of Pupils

In addition to directing "learning" one other very important task of the teacher is to promote personal and social adjustment of children. The majority of children behave normally and are well adjusted in life situations. But there are a few who deviate from normal course of behaviour and are maladjusted in different degrees. Some of these children misbehave with the teacher and show off in the class. Some others steal or cheat, are lazy or unsocial and withdrawing. And still some others are truants or deface the walls, use foul language and break the light posts. These are *problem children.* These children need sympathetic help from the teacher. The causes at the root of such deviations in behaviour have to be found out and removed. Most of these behaviourism result on account of the child feeling frustrated in personal and social relationships. Teachers should assist these children to set realistic goals which can be achieved. The need for keeping discipline will diminish when children are happy and are busily engaged in activities that appear to them as being of real worth.

Teachers would find a lot of material in Educational Psychology on the basic processes of adjustment. Educational psychologists have borrowed a large amount of material concerning children with problems which handicap them in learning, from the field of mental hygiene, clinical psychology and social psychology. Educational psychology is also concerned with those family, community and cultural influences that affect learning and adjustment. A teacher who seeks the help of educational psychology in solving problems in learning and adjustment shall find that information from these fields has thrown considerable light on the social dynamics of the school and the classroom. Educational psychology furnishes the teacher with the necessary facts and techniques for analysing behaviour—both this own and that of others—to the end that normal adjustment may be facilitated and growth in personality effected.

Evaluation of pupil qualities and characteristics using educational and psychological tests helps the teacher in *effective guidance and counselling*. For instance, if the achievement test is used sufficiently early, difficulties can be determined and remedial work undertaken to help the student. The test results, presented in the form of a profile chart, help the student determine his strengths and weaknesses, thereby furnishing him with useful and dependable data for making educational and vocational decisions. "Guidance of all kinds has a common purpose—to assist the individual to make wise choices, adjustments, and interpretations in connection with critical situations in his life. This is done in general, through (1) information that he is helped to secure, (2) habits, techniques, attitudes, ideals and interests that he is helped to develop, and (3) wise counsel by which direct assistance is given him to make the choices, adjustments and interpretations." Guidance and counseling go hand in hand though the former is more general and the latter more specific. Individual counselling is a vital part of the guidance programme of the school. Students have difficulty with educational problems involving learning and decision making. They have problems that are primarily emotional in nature and problems that are primarily vocational in nature. The teacher has to help them solve these problems. He has to act as their "friend, philosopher and guide."

"Individual counselling is communication between counsellor and counsellee relative to the situation of the latter. Such communication has as its primary purposes the enabling of the counsellee better to cope with his immediate and specific difficulties and to develop his capacity for self direction and self understanding in order to meet more effectively the problems of living in the future." As was state earlier, the teacher's position is similar to that of the medical practitioner who must first study his patient before prescribing for him. For this purpose, the teacher should have a good knowledge of evaluation procedures and use of evaluation results, which he is not likely to have without a sound knowledge of educational psychology. A vital part of educational psychology is "Measurement and Evaluation in Teaching," which would help the teacher in using the results of evaluation procedures for guidance and counselling. "Assisting a pupil with educational and vocational decisions, guiding him in the selection of curricular and extracurricular activities, and helping him solve personal and social adjustment problems, all require an objective knowledge of the pupil's abilities, interests, attitudes and other personal characteristics. The more comprehensive the picture of the pupil's strengths and limitations in various areas, the greater the likelihood of effective guidance and counseling."

Educational psychology has something significant to say about the conflicting roles, adjustment problems and mental hygiene problems of the teacher himself. It can help teachers solve personal and emotional problems. Teachers, although relatively stable when compared with other professional groups, need assistance in improving their mental health. This need takes on added importance when the undesirable effects upon pupils of teacher maladjustment are considered.

The teacher is a *professional person*. Unless he is in possession of a body of basic knowledge, basic principles and basic laws that are the products of scientific research and scholarly activity in educational psychology, he cannot operate with professional skill and exercise professional judgement. In the teaching enterprise, besides the teacher, there are *educational administrators*. To these administrators, the values of educational psychology *belong to* in one way or another just as they do to the teacher. In many ways they have to make wise decisions that influence the pupils, school and community. They should be able to judge the extent to which the objectives of the school are being achieved, to identify strengths and weaknesses in the curriculum, and to appraise special programmes in the schools. They should have the required information on which to base administrative decisions concerning the placement, grouping and placement of pupils. They should collect objective data to be used in explaining to the community the goals and accomplishments of the school. Learning how to make judgements and to take decisive action, plantfully and carefully, require the administrators and teachers to understand themselves, their pupils and the processes of learning and intellectual development. These understandings are important parts of educational psychology.

Limitations of Educational Psychology

Educational Psychology arise from the nature of "Teaching", the nature of "Science", and the nature of "Psychology" itself. Teaching is more an art than an applied science. Therefore, books and lectures and other media of information render us relatively little help. The only choice left to us is to depend upon the natural aptitude and experience of the teacher. Between two engineers, one old and the other young, you may choose the young engineer on the ground that he is well versed in the modern techniques of engineering. But in so far as music is concerned, that particular musician who sings the song you desire to listen to is needed. Very often science is indispensable in the solution of a problem. Yet, there are problems which no scientific information in the world can help us to solve.

These problems are related to philosophical, ethical and

religious values. To make any decision, we have to consult them. Educational psychology has all the limitations of science. It has a good many limitations on its own account too, that some other sciences do not have. Psychology is not as yet a highly integrated science. It possesses a vast number of facts like most of the other sciences. But unlike many of the other sciences, it has not yet been able to integrate this body of facts into any closely knit pattern. "Thus, whereas the student of applied physics learns relatively a few general laws from which a myriad of specific facts may be deduced, the student of applied psychology must be prepared to get the answer to his problem by consulting spefcific investigations bearing on these problems." These possible limitations should not discourage us. If teaching is an art, teachers with a natural aptitude, and experience, and capable of personal supervision should be preferred to teachers who possess a body of information only. If science does not answer all questions, we must be prepared to deal with values, even though they may be questions of philosophy, ethics or religion, and exercise shrewd judgement.

4

CHILD NATURE PLACE AND ORIGIN

The child guidance movement started formally n this country about fifty years ago. During this half century there have been many changes not only in the child guidance clinics but also the professions working in these clinics. Child psychiatry, which did not exist at that time, has gradually accumulated a body of knowledge and skills of its own so that it is now recognized as separate special field within psychiatry.

The children with marked anti-social behaviour are categorized as juvenile delinquents. Broadly a juvenile delinquent is a child generally between the ages of seven to eighteen years and whose anti-social tendencies are so grave that they become or ought to become the subject of official action. Delinquent behaviour is a rebellion and an expression of aggression which is aimed at destroying, breaking down or changing the environment. The definition of a delinquent various according to community attitudes, the state of law and the policies of law enforcement agencies and the kind of behaviour which led to a delinquency level depends upon the kind of society in which one lives. "Juvenile deviance has been an important issue in every society in time and space. Therefore this phenomenon has been giving rise to extremes of emotions in affected people and non-conforming act of a juvenile has been evoking mixed reactions against concerned depending upon the area, cultural norms, social attitudes, position, placement, relationship with the perpetrator or victim, responsibility, interest, knowledge, discipline, orientation and outlook of the people defining and applying social standards on the activities of the juvenile."

Indian society has achieved considerable progress and development in different spheres of social activity. But this rapid advancement has also brought many socialevils in its stride. Indulgence in anti social activities by the students at a very delicate (adolescent) age is one of the major evils. The principles of secondary schools have confronted with problems pertaining to anti-social activities like, gambling, stealing and drug abuse among the students of their schools. The speed with which such behaviour is spreading among the student community is the most dangerous aspect of it. Future planners and administrators of schools are in general worried about the intensity/magnitude of the problem. The resettlement colony in question, is generally inhabited b the people of the lowest strata of society.

It is also feared that in such colonies anti-social elements like gamblers, pick-pockets, drug addicts etc. lie in good numbers. The students coming from such homes and surroundings are likely to be more prone to these anti-social activities as compared to the students residing in other colonies. The problem does not concern schools only. Seen from a wider angle, it is a social problem which concerns everyone in the society. But school is supposed to be an institution where there children are taught about the good and had behaviours. More so to the children coming from parents who have not been able to set a model to their children. These children are a part of the society and the country is equally dependent upon this section of the population. Such problematic children may prove to be cancerous to the growth of the society and weaken the foundations of the countries future. It is a sheer waste of humanresources and miscues of manpower. In school, it leads to other serious problems like breaking school rules, damaging school property, redeness and misbehaviour with teachers etc. The net result is the entire atmosphere of the school is spoiled and other innocent students may suffer the consequences. The problem in its serious form does not spare the home as well. The parents are worried over the behaviour of their sons when they (sons) return late at night, use abusive and filthy language, are caught eve-teaching stealing, gambling, pick pocketing and indulging in other violent activities. It is a disgrace to the society and demoralizing for the nation when children start indulging in anti-social activities and criminally waste their talents and energies by putting it in undesirable directions. The school as a temple of learning and the teacher as a nation builder, is duty bound to use his influence and act in a manner which not only prevents the problem from originating but also checks and cures it and enables the spoiled children to adjust in the new circumstances. This is as matter of fact, progressive function of education which ultimately results in national strength and social uplift. It is therefore, imperative to examine the factors responsible for such anti-social behaviour and to find out possible and suitable remedies for quick solution of the problem.

The major objectives of the study are: (a) To identify the nature and magnitude of the problem, (b) to analyse the major causes of the problem, and (c) to suggest suitable remedies for overcoming the problem in schools.

The study covers only few anti-social activities such as stealing, gambling drug abuse and smoking reflecting anti-social behaviour. Except smoking the other three activities are considered as of anti-social nature form legal point of view also. But the investigator includes smoking also as an anti-social activity on the following grounds.

During this time much has happened in analytic work to throw added light on the problems concerned and to influence the conditions of life for young people, whether normal or abnormal, Nevertheless, in spite of partial advances, the position with regard to the analytic study of adolescence is not a happy one, and especially unsatisfactory when compared with that of early childhood. With the latter period, we feel sure of our ground, and in possession of a wealth of materials and information which enables us to assume authority and apply analytic findings to the practical problems of upbringing. When it comes to adolescence, we feel hesitant and, accordingly, cannot satisfy the parents or educational workers who apply for help to us and to our knowledge. One can hear it said frequently that adolescent is a neglected period, a stepchild where analytic thinking is concerned.

Place of Origin of the Problem

The parents have stated that their children had acquired the bad habits at the school in the company of their friends. The place of originating these habits is not the school but the home and social surrounding in which the child lives and moves and spends about 80 per cent of his total time. It can be noted form that 100 per cent cases of stealing. 80 per cent of gambling, 75 per cent of smoking and 20 per cent of drug addiction originated from society and percentages in case of home for the above cases are zero, 20, 25 and 80 respectively, whereas none from the school campus.

It is, therefore, established that home and society are equally responsible in generating these bad habits and are the breeding grounds of anti-social behaviour. The cases investigated under the present study ranged in the age group of 15 to 19 years and the age of origin of the problem ranged between 11 to 17 years.

Nature of the Problem

The problem identified under anti-social behaviour were mainly four in nature. They are smoking, drug addition, gambling and stealing. Smoking leads to drug addiction followed by other antisocial activities like gambling and stealing. When the smokers fall into the trap of drug addicts, they first start with a very small quantity of the substance placed inside the cigarettes etc. After sometime they start enjoying the 'kick' and need more. Since the drugs are very costly and beyond their reach, they start stealing money and materials from home to satisfy appetite for drugs. Very soon they learn the art of stealing and start laying their hands over other people's property and pockets. Money earned through this way is misused for gambling in the greed of earning more money and the circle is complete. Once they become addicts or slaves of the drugs, they cannot live without it. Then in the absence of drugs they feel severe pan in the stomach and the whole body, and the eyes stat lacrimating and nose starts flowing. The adverse effect of the drugs surfaces very soon. The fingers get burnt, black spots appear around the eye lids, the body becomes while and the child looses all interest in studies, games or any other work. It is estimated that there are eleven lack drug addicts in Delhi alone. This has created greater risk for our children who may fall victim to this evil habit in the company of such people.

Causes of the Problem

According to their higher percentages some of the factors can be put in the category of major cause. They are bad social surroundings, lack of EVGC facilities, bad habits of family members, bad company, illiterate parents, poor living conditions at home, lack of child's interesting studies, and lack of activities in the school.

The knowledge of these causes can prove of immences help in finding out suitable remedies.

The various cases of the problem have been analysed under the following broad categories:

(a) Individual and Behavioural, (b) Home, (c) Social (d), School.

The child himself with his attitudes and behaviour is responsible for learning bad habits to certain extent. The that child's lack of interest in studies contributes to the problem to the extent of 65 per cent, his disobedience, unruly behaviour and in disciplined attitude to 25 per cent and poor health to 15 per cent. The percentages mentioned above do not suggest any great intensity in this particular cause. This means this cause can be effectively routed out by taking suitable remedial steps.

This dimension examines the living conditions at home in terms of housing accommodation, size of the family, education of the family, income of the family, habits of the family members and other facilities at home. Looking at the large size of family members ranging from 5 to 9 with an average of 7 the accommodations in respect of 90 per cent families are badly overcrowded and insufficient.

In addition to all this, the behaivour and habits of the family members do not seem to be condusive to the proper formation of habits of the child. It can be clearly seen from that 85 per cent fathers and 20 per cent clearly seen from that 85 per cent fathers and 20 per cent mothers are smokers, 45 per cent fathers and 5 percent mothers are drug addicts, 5 per cent fathers, and 5 per cent mothers are gamblers while 5 per cent fathers and 5 per cent fathers and 5 percent mothers are gamblers while 5 per cent fathers and 10 per cent mothers indulge in stealing. In short a large number of houses lack minimum facilities, are over crowded and their earnings are meager that means poor living conditions are found at home.

The behaviour of the child already infected with such habits produces tremendous influence over the behaviour of other innocent children around them. The study shows that some of the cases acquired these habits at the school possibly because of lack of spare time and also non-availability of safe and lonely place - which are essential requirements at the initiation stage.

However, when the same children move in their social surroundings they come into contact with other children while playing or simply loitering around. During the course of investigation undesirable social surroundings and bad company have been found to be the major causative factors responsible to the extent of 95 and 85 percent, to the problems respectively. The two are also interdependent and supplementary teach other. For example, bad company produces bad environment (or bad social surroundings) and in the same way bad social surroundings lead to bad company. The children there, naturally, move around for playing and other purposes.

Some of them due to individual, home or school problems are easily attracted towards other children or even grownups, who have already acquired bad habits, and are thus initiated into such habits and anti-social behaviours. Such habits have invariably started with smoking in the first place. It can, therefore, be concluded that the influence of societal factors, such as bad social surroundings and bad company, are more pronounced over the behaviour of the children than any other factor.

The school has not been found directly responsible in originating the problem. However, it has been causing and contributing to the problem to some extent by failing to take preventive measures and also by creating situations under which the child becomes vulnerable to fall prey to these problems very easily. That means had there been these facilities a good number of students could have been prevented from falling in to the trap of bad habits right at the beginning. In the same way lack of activities created scope to the extend of 40 per cent while rudeness and indifference of teachers produced negligible effect i.e. only 5 per cent. Lack of interest in studies has been responsible in causing the problem to the extent of 65 per cent. Lack of interest in studies has been responsible in causing the problem to the extent of 65 per cent. Though it is an individual factor but the schools are also responsible for this individual indifference to the extent it was caused due to bad teaching and lack of other facilities in the school. This view has also been and lack of other facilities in the school. This view has also been confirmed by a recent study on anti-social behaviour which concludes thus, "the teachers and people in management themselves might be poor in prosocial value and behavioural orientations. They might be loving to terrorise the students by enjoying dominant position entrusted to them by the society and by getting its support."

Recommendations

The problem of anti-social behaviour of the children has been the cause of concern of not only the parents but also many other professional groups. The educators, administrators, clinical Psychologists, guidance counselors, social workers, sociologists, psychiatrists, the criminologists, the police and judges and at one time or other it has "captured the attention of many different types of citizens from the man on he street to the president.."

There are a number of child welfare institutions functioning in Delhi correcting/curing and reforming the problem of children and putting them on the right path. They swing into action only when the child commit some criminal offence i.e. when he is declared a proved juvenile delinquent. Barring a few, no organized and systematic effort has been made by any agency or institution to go deep into the causes of the problem and take steps to prevent the problem of occurrence and to check and control it, if it is already in the process of growth in certain children. Only the preventive measures supported b curative efforts may succeed in routing out the problem from our society. This preventive programme starts at home, goes to the neighbourhood, extends to the school and covers all other community institutional structures.

Here the role of individuals and institutions/agencies has been examined with a view to prevent, control and eradicate the problem of anti-social behaviours of the secondary school children:

- Parental Role
- Role of the School
- Role of the Teachers
- The School Counsellors
- The Police

Parental Role

It is incumbent on the parents to set an example of good behaviour at all times to children since a child draws his main identification from the parents he loves and regards best. Responsibility for the children vests with both mother and father. It would therefore be appropriate to identify the major roles of the parents, who even by their mere presence at vicinity – or distance through their latent or indirect support to the child – provide a protective umbrella thereby still a sense of security and confidence in a child. This reality plays an over bearing role in shaping and providing direction to the personality of the child.

The parent should show discipline by their own conduct. Mother and father should always present a united picture. They are expected to act as models not only in regard to ideals but also in terms of responses and reaction patterns of varies life situations and for this first of all they should give up their own bad habits.

Appropriate status should be accorded to the child among the members of the family. This satisfies adolescent's instinct of recognition and develops qualities like obedience, feeling of responsibility etc. in him.

Any reward or punishment given to the child should be

justified and the purpose properly understood by the child to avoid any kind of malice from the parent-child relationship.

According to his age, potentiality and reflexes, the need, importance and efficiency of social values and norms along with roles and responsibilities should be explained.

Proper supervision and control should be exercised over the child - to have a close watch at his activities, interests, company of friends, his movements etc. and apply preventive measures wherever necessary without putting unnecessary check on his freedom.

The basic requirements of the children should be fulfilled as far a possible.

They should be treated with due sympathy, affection and consideration.

The parents should keep in constant touch with the school and the teachers in order to keep themselves abreast of the progress of the child and also to find out solution of various problems facing the child.

But the parents of the colony under consideration are mostly illiterate and belong to a very low income category, therefore, they may not be expected to follow all the suggestions given above. Under the circumstances the responsibility of other agencies (outside home) becomes greater in dealing with the problem and the school in particular is supposed to play greater role in this direction.

School

School has not been found directly contributing to the problem yet it can play a very constructive role in the prevention and control of anti-social behaviour of the children. The school may succeed in their efforts in this direction only when they become good schools and are able to coordinate their efforts within the framework of total community endeavour. The school should create favourable condition for attracting and creating interest amongst the children and for the balanced growth and development of their personality by observing the following suggestions:

Should provide all necessary facilities which help in the all round development of the child.

Should differentiate instructions to suit the interest and abilities of the problem of child in order to ensure some measure of success and notification for them.

Should provide special services such as remedial instructions, counseling, testing, medical examination, home visitation case work, special health services etc.

Should maintain reasonable class and school size say for example, aroudn30 students per class and 500 per school.

should provide opportunities for school - community interaction by organising P.T.A., open day for the parents and society, playing matches between school and community inviting community members to participate in various activities of the school and vice-versa. Closer cooperation between parents and teachers will certainly help in moulding the children on correct lines.

Should identify the problem children through cumulative school records, tests, medical examinations, mutual confidence, home visits etc. and should accept them as a person as they are normal human begins though victims of circumstances.

However, the problem can not be solved by the school alone or for that matter by any single agency - be it home, clinic, or any other institution. The problem is of complex nature and calls for coordinated efforts of all community forces of which the school represents an important segment.

Adolescence is a period of extreme in behaviour.

Emotions of adolescents are intense in nature. On the basis of emotions an individual develops sentiments, values of life, ideals and virtues, which are the basic ingredients of character. This is the period when child is very much vulnerable and receptive to any kind of good or unhealthy effect/behaviour. The teachers' proper guidance and control is very necessary during this period.

"The teacher should organise and control emotions of an adolescent from the very beginning instead of repressing them. He should take care to explain the real situation to an adolescent and create a better environment avoiding all sorts of strains and mental conflicts otherwise he will develop a particular temper, movement, posture or clumsiness. Teacher should be on guard in relation to this type of children and treat them sympathetically.

The teacher should:

– Aim to develop self imposed control within the child

– Counsel problem children regarding their educational, vocational and personal problems

– Identify problem - prove youngster and refer them to E.V.G.C. of the school.

– Develop and use cumulative education and other records.

– Help the problem children from time to time, to appraise his own growth and achievement, providing thereby a sense of security through success.

– Have working knowledge of child's physical, emotional and mental growth and development which will enable him to collaborate with remedial, psychological and social work professions

– Present an acceptable attitude towards the problem child and also show an effective example for imitation through is behaviour habits and emotional maturity.

– Make effective use of knowledge of the basic human, physical and psychological needs and to establish specific and meaningful goals within the range of child potential.

The role of school counsellors is being progressively recognized in the achievement of individual educational excellence. In America the "Policy makers and special interest groups recommended an expansion of such services, a sharpening of counselors roles and planned an interaction between school counselors and parents, teachers and others as well as setting priorities in how counsellor's tie and skill will be used in a particular school and with particular group of students.

The counselors may prove of immence help to the school in particular and society in general by providing necessary guidance and help to the needy children, who are facing the problem or are prone to the problem in the following manner:

* He should keep a close track on the activities of such children, on daily basis, in all classes when the child moves from one class to another through Daily progress report (DPR) and case study.
* Should identify and teat drug dependency, aggressive behaviour and other bad habits.
* Should enable the student to cope with a wide range of development and social dilemmas.
* Should help the students to plan, choose and succeed in vocational education and in higher education.
* The counsellor's functions in general should include cooperative activity with teachers, the use of group guidance techniques, computer assisted guidance, comprehensive career information system and related

methods designed to provide career guidance to all students.

The National Commission on Secondary Education (1984) held the view that the school counselors were important to excellence in vocational education.

In India though he importance of the services of councellors in the field of education is well organized but the face of expansion of school guidance and counselling service far from satisfaction. For example, out of the five schools selected for the present study educational and vocational guidance counselor was found in one school only.

A survey conducted by Perin H. Memta reveals that only 55.28 per cent (i.e. 504 schools were covered by some kind of guidance service in 1981 and that only 78 counsellors were diploma holders in guidance. Some urgent steps should therefore be taken to provide efficient guidance service in all the schools.

- The counselling function should be discharged by professionally trained counselors.
- The counsellor-student ratio should be about 250 students for counselor, so that they could discharge their duties effectively.
- A separate guidance and counselling room along with all the necessary facilities should be provided to the counselors in every school.
- Relevant literature should be made available to the counselor for his/her small library and proper display on the wall boards of the guidance room.
- Heads of the schools should also be properly oriented about the guidance services, so that they may properly supervise and guide the counselors and may not assign them any other work of the school apart from

the work for which they are meant. In other words, they should not be expected to do everything.

Keeping in view the importance of counseling service with regard to the benefit of school children in general and problem children in particular, a clear-cut long term policy for providing guidance serving and provision of sufficient funds by the Govt. for implementing the policy are essential prerequisites. One healthy and happy child will make a healthy and happy nation and no other charge of a Government can be more important than to take care of its children.

The police plays an important role in all the stages of behaviour modification of a delinquent, i.e. during location apprehension, custody, prosecution, treatment and rehabilitation. But their role at the stage of prevention is crucial.

The present study bad social surroundings and bad company have been found as major factors responsible for producing and perpetuating anti-social behaivour amongst the children. Here the police may prove of much help in preventing and controlling the problem directly as well as indirectly - by cleaning all vice-dens. This is very easily possible for then as they have vast resources, sufficient knowledge on the subject of crime and delinquency, reliable information about delinquents as individual, greater exposure and association with the public etc. to accomplish the task.

The anti-social child does not constitute a unique type of human being. They have the same human desires and natural yearnings as anyone else. They are also human beings though the victims of faulty upbringing and poor school and social environment. Therefore, to prevent the children from failing into as such habits and to control and cure those who have already caught into this phenomena, some remedies have been suggested/recommended under this study. In the first place parents should try to provide a congenial atmosphere at home for the child's study and relaxation/recreation, give him his rightful place in the family, show sympathy, love and favourable attitude, provide sense of security, fulfill his basis requirements, make proper supervision along with necessary freedom to the child and keep constantly in touch with the school. The role of school and teachers is equally important in tackling the problem. The school should try to create an environment in which the child can be kept usefully busy and provide all the necessary facilities to him, according to his interest and aptitude, which are helpful in the all round development of his personality. It should seek help and co-operation from the parents and the community.

The teachers command tremendous regard in the heart of students. They should know every child in their classes and should teach and help them wit full sincerity, devotion and affection. Their own attitude, behaviour and personality, should present a model of virtues for the child to emulate. They are really capable of doing miracles if they are willing to do so. The school counellors are trained lots in tackling the problem-children. They should follow a treatment procedure which fits in individual offender and not merely the offence, for which knowledge of working of the mind of the individual offender is very essential. The school counselor can prove to be of immense help in shaping the behaviour of problem children; but still they are found in a very few schools and there too they lack basic facilities for performing their jobs efficiently and skillfully. The education department is, therefore, advised to give this job top priority and allocate necessary funds to post such counselors in all the schools.

The police can also play a very important role in prevention, identification, treatment and rehabilitation of problem children. It can help in improving he social-environment and removing/reforming bad characters-which are the major factors causing the problem of anti-social behaviour. The police should, therefore, be given suitable training in order to bring a change in their attitude and to enable them to do this social defence work with full cooperation from the society and others engaged in this work. Any programme which aims to identify the potential problem at an early age, which tries to study and diagnose needs of the child, before and after the problem and which aims to provide treatment suited to individual diagnosis, may be expected to succeed to a great extent. But this is only possible when all the agencies/institutions and individuals concerned with the welfare of the child make a well knitted and coordinated effort in this direction. The problem of anti-social behaviour amongst the students cannot be successfully tackled unless thorough investigation is made into the causes, leading to this evil not only in resettlement colonies but also in different settlements inhabited by the people of varied socio-economic status.

5

CAUSES AND CONDITIONS OF PSYCHOLOGICAL ETIOLOGY

The Alexander and other similar psychiatric classifications limit the dynamics of law violation too narrowly in focusing attention solely upon the neurotic offender. Various other relationships of the individual's subjective states to his misconduct will be considered below. First, however, the neurotic delinquent should be examined more closely. The neurotic offender is usually the product of an atmosphere of over rigid parental restraint, of overprotection and over restriction, frequently of perfectionism and inconsistency; the child's ego lacks the strength to sublimate his impulses into socially acceptable channels. But instead of attempting complete repression, which often leads to ordinary neurotic symptoms of hysteria (such as functional blindness, deafness, lameness, etc). Anxiety, obsessions, or compulsions, the individual here commits offenses which offer release to his only partly inhibited drives. Neurotic delinquency may, as Freud suggested, develop out of a sense of guilt in reference to socially unsanctioned wishes (Oedipal, Homosexual, or heterosexual); the individual may be seeking punishment to dissolve his guilt feelings. Where the impulses are wholly unacceptable to the ego (rather frequently in the homosexual personality, for example), a substituted expression-often itself painfully, repugnant-may be permitted as an unconscious preference. Thus such compulsive offenses as exhibitionism (indecent exposure), voyeurism (peeping), kleptomania (compulsive theft), pyromania (compulsive arson), extreme sadism (erotic pleasure in inflicting pain upon any object), masochism (Pleasure in receiving pain), or other sexually expressive behaviour may reveal the individual's repressed and pervertedly redirected drives). The following cases are illustrative:

The case of Albert Thornley, studied intensively by Ben Karpman and presented in detail from the records of his psychoanalytic therapy, illustrates very well the behavior dynamics in a neutotic individual "entangled in a maze of complicated mental states from which criminally sometimes offers an immediate, albeit wholly unfit, means of escape."

Thornley had an extended criminal history, including theft, extortion, fraudulent enlistment in the army, repeated drunkenness, and many minor offenses, though these were only a continuation of the delinquencies which developed in his youth. His problems of adjustment were mainly a product of the distortions in his intrafamilial relationships, serious enough to a lead to series of commitments to psychopathic institutions. A combination of an excessively close attachment to the mother and hate of the father wee at the root of his psychoneurotic and psychopathic trends. Through pout the history one finds repeated evidence of the patient's remarkable emotional attachment to the other, an attachment which influenced much of his subsequent behaviour, as his penchant in later years to associate with women old enough to be his mother. He displayed some insight into this fixation: "A short time after I finished school, my mother died, and when she left us it seemed as though something had gone out o my life and nothing has ever come into if to fill the empty space. Like most boys, I had grown to depend on her for advice; but, unlike many bys, I was more than willing to do what she asked e to do, and perhaps, I Loved her too well." Of his relationship to his father, Thronley cites a traumatic instance:" My parents once had a family quarrel and I ran out of the house and told the neighbors all about it.. After father's return from work, we were sitting down to the evening meal when mother told him what I had done. He said nothing to me during rest of the meal: but immediately afterwards told me to remove my clothes and then tied me to an old bed and started to whip, I have been whipped many times since then, but never like this. Can one imagine how it seemed to a boy scarcely more than a baby to be whipped by one be should love instead of hate? Little did I know that of all the less I have been taught, this one would stay with me long after the other had been forgotten. After this whipping my back was sore for a week, but strange to relate I did not run to my mother when I came downstairs down stairs; I went to a corner, sulked for awhile, and then went to bed where I cried myself asleep."

When Thornley was arrested for a minor offense his father recommended his reform-school commitment; there he learned all he did not already know about crime and sex. In the ensuing years his emotional dependence on the mother he had lost and his psychologically incestuous sexual relationships with older women caused him much pain and anxiety.

"In later years I regretted this trait in my disposition, but I honestly can't help it. I have always felt that a woman with whom I have had intercourse was not worthy of any man's consideration. From this you can see how much I loved my mother, both while she was living and after her death. His own marriage was rather naturally an incompatible failure and contributed to his neurotic escapes. The main reason for his social conflicts appears to have been the degradation of his personality consequent upon the breaking up of the home, allowing him to sink lower and lower in his industrial adjustment, often reaching an impasse at which some sort of antisocial activity seemed to him the only avenue of escape." The compulsive character of the illegal acts committed by some neurotic offenders, particularly where there is a distorted psycho-sexual history, is illustrated by the following excerpt from the public press:"

A lovesick young man, who lit fires to ease the loneliness after bidding his girl good night, was arrested in Newark early yesterday while kindling a flame in the cellar of a Forest Hill home.

Jersey police charged the prisoner, identified as J-S-, with arson, According to police, S-told of having started "at least four fires in Newark since November I." Assistant Police Inspector John Dolan estimated the damage at $20,000. S-was caught starting a blaze in the basement of [a building] by a tenant who was returning to his home after watching another fir ea block away.

The amorous arsonist told the cops he always got lonely after saying good night to his girl friend. He would yearn for some excitement and he'd start the fires to ease the strain and warm himself up inside, the police said. The following case illustrates very well the consequences of an over-rigid training in a "good home" that resulted in a high degree of moral anxiety:

Years ago I was consulted about making plans for a six-year-old girl because she insisted upon being skipped to a higher grade in school although she had started school earlier than the average child. She seemed bored with the low educational standards of the class, and she showed signs of severe social maladjustment. She was born in luxury, the only child of a foreign diplomat. Although the parents had continual social engagements, the mother tried very hard to give the child all the care that she needed. She had a nurse who surrounded the child with loving care and attention. The mother, who was a very intelligent and beautiful woman, emphasized that the child never had given any trouble up to the time of consultation. In fact, Bardber had been "a lady" almost since birth. She never had to be reprimanded, was always compliant. Thee never was a commotion in the nursery. Barbara never indulged in any childish mischief; she was always serious. It was not necessary, the mother stated, to punish Barbara because she always did every thing perfectly. If her mother or her nurse, both of whom she seemed to adore, explained it to her, she would accept everything without opposition. I never could find out how this smooth development could have been brought about in the earliest ego development, when verbal understanding is not yet possible. Indication of deep-seated emotional antagonism between child and mother seemed to have started in the form of a incompatibility at breast feeding. This was both unsuccessful and tormenting to mother and child. Nevertheless the mother continued her deliberate effort to be a model mother. Although Barbara never liked to eat, she made a great effort to consume large amounts of delicious food. If the food were removed, she would be discouraged because she felt people would be dissatisfied if she did not eat it. She was never really forced to eat in an overtly aggressive way, but she was told that she would growing and strong, and wasn't tat what she wanted? In Barbabara's case the unnaturally overprotected, exacting, and polite atmosphere from which aggression in its overt manifestations was completely abolished, compelled the child at a very early age to take, without overt anger or protest, the attitudes and unspoken commands of her environment (in her superego). Thus she had no opportunity to express her aggression and to release her tension in motor manifestations.

Barbara was forced to repress her aggression early and completely, whereas children need a chance to obtain control over their aggression and to repress the motor manifestations gradually and piecemeal. Barbara could not let herself be aggressive, because everybody around her was soft-spoken and mild. The danger of being diametrically different from the environment made her feel to keenly that impulsive behaviour would case loss of love...the ego-ideal erected was beyond what her comparatively weak ego could ever truly hope to live up to. Her ego was further weakened because, in order to keep in check such strong and abruptly executed repressions, she had to put up strong secondary defenses, so-called "reaction formations." She had to be a model child, a little lady. All that was at the expense of her ego, because the psychic energy in such cases is used up n keeping up the defenses instead of being utilized for constructive active functions. Psychoanalysts call this process the impoverishment of the ego...Barbara's ambition was infinite, but ambition is not ego strength. Such ambition is the superego' expectation that the ego achieve miracles. Such an ego gets continually discouraged; the gap between the ideal and the capacity is very wide; the discrepancy is felt as an inner sense of discouragement, dissatisfaction with oneself. This is the anxiety, which Freud called moral anxiety, that is felt as conscious or unconscious guilt. Superego anxiety, in fact, is that type of inner far that characterizes most children with primary behaviour disorders of the neurotic type.

The Psychopathic Delinquent

The Alexander and Staub schema too narrowly limits the varieties of offenders with a psychological eatiology, for there is no doubt that other types of psychogenic influences deserve a more important position that those promoting neurotic delinquency. One additional variety that has been prominent in recent psychiatric and psychoanalytical investigations is the psychopathic personality. The psychopath has been interpreted is the psychopathic personality. The psychopath has been interpreted as definition inhibition through the absence of an effective, coherent superego. He is a most unsocialized aggressive child-cruel, defiant suspicious, egocentric, unfriendly, vindictive, primitive, and hypersexual (or sexually uncontrolled). In extreme manifestations, he shows an utter lack of capacity for sympathy; he is quite without compunction for his ruthless exploitations and injuries to others. Such a person usually comes from a home situation defective in affection and control. He is an unwanted child, unloved, consistently rejected; his response is uninhibited hostility and overt aggressions. Characteristically a deprived person, his most serious deprivations are emotional. Because of the seriousness of his misbehaviour and his lack of identification with others, the psychopath is a difficult problem in treatment; the therapist requires infinite patience to provide him the consistent. Close, and warm understanding a that he needs. His history of rebellious responses to chronic rejection makes it almost impossible to come into touch with him. Usually committed to a disciplinary institution, the psychopath merely becomes more embittered and callous. Even if he receives therapeutic attention, the structure of his personality has ordinarily been too rigidly fixed from his earliest days to achieve much success with him.

Among various writers it has been customary to apply the term "conduct disorders' to the sort of symptomatology found in the psychopathic personality, especially in relation to children. John Van Ophuijsenis quoted in Lewis and Paccella as follows:

It is customary to speak of a conduct disorder whenever there is a deviation from the accepted code of morals. This is why conduct disturbances are also referred to as character disorders; the term, however, includes the "bad" habits and should not be substituted. In accordance with the before-mentioned criterion we find listed as conduct disturbances; lying, truanting, stealing, disobedience, running away, destructiveness, fighting, sexual activities, etc. In typical cases one or more of these disturbances are mentioned as the reason why parents or their substitutes or school authorities seek the help of the psychiatrist or of a child-guidance clinic. The invariably add information to the effect that the child is disobedient, unmanageable, and disrespectful and that the does not respond to punishment or show any sing of guilt. In many cases they also add tat the difficulties started at an early age, at three years of even before. "The child was always bad." Even at that time there was marked disobedience and rebelliousness; even then they had noticed the child's tendency to run away from home, or to lie, or to steal (e.g., pennies from the mother). Sometimes they mention that the child was not punished at first or at least very little, but the story always ends similarly; even severe physical punishment produces no improvement at all...

In the mind of the expert the constant conflict with the environment and the absence of guilt, which characterize every single typical case, very soon begins to appear as two sides of one abnormality. This abnormality seems to lie in what has happened to the child's aggressiveness: he shows much more aggressiveness than the average child in his contact with the environment, i.e., much more external aggressiveness, and he shows much les aggressiveness than the average child in his attitude toward himself, i.e., much less self criticism and reproach, which are the expressions of internalized aggressiveness,. Correlated with this and related to it, the examiner finds a narcissistic self-evaluation. The child appears to think of himself as come thing exceptional, great, strong, a "Big sot," etc., always with the expectation of being able to accomplish his aims regardless of the impediments formed by laws, customs, and moral code.

Abnormal aggressiveness, absence for defective development of guilt feeling, and narcissistic self-evaluation form the triad that characterizes the child with conduct disturbances.

A further contribution to the theory of the psychopath has been made by David M. Levy n his study of types of maternal overprotection, wherein he divides psychopaths into two groups, he indulged and the deprived, each with its somewhat distinct characteristics. It is generally agreed that the parent's rejection of his child is the basis either of indulgence or deprivation, though in the former mechanism, the guilt feelings experienced by the parent produce exaggerated and sporadic reactive efforts to display love. Levy points out that indulged children, undisciplined b their mothers, display infantile demands and expectations, but they are not characterized by emotional shallowness. They are capable of forming warm human relationships, although exploitation is the usual consequence of these relationships with the over protected. The deprived psychopath, on the other hand, is unable to establish identifications with person in his environment because of the emotional coldness that marked his early life. In the absence of sympathetic relationships the psychopath fails to develop a conscience or a capacity for warmth.

The case of Isaac represents the pattern of the psychopath in his typical manifestations; he differs from others mainly in having responded to treatment:

Issac, a boy of fourteen and one-half, with high intelligence, was committed for stealing, truancy, vandalism, sexual perversions and general incorrigibility. He was a handsome boy who always made a good impression until people got to know him; and only child whose mother died and whose father abandoned him when he was two years old. Reluctantly cared for by relatives, he as passed from one branch of the family to another, and finally to an orphan's home. From his very infancy he presented serious behavioural disorders.

In the institution he had no regard for routine or discipline; he made the most unreasonable demands. He was bitter and antagonistic, even to two people whom he processed to like the farmer and his social worker. His pleasure in farm work was encouraged even at the expense of formal school work in which he had no interest. Both these people spent a great deal of time with him remained consistently patient and sympathetic, and made no comment on his general misbehaviour. His poor behavioural was discussed with him when it was directed against these two people. They would stress their desire to continue to be his friends, but would point out that behavioural such as this provoked others into disliking him. Continuous exposure to this kind of handling brought out a good deal of self-blame on his part, and some expression of doubts as to his worth-whileness. He hung to these two contacts and made conscious efforts to see their, advice and to follow their guidance. As he was able to achieve some success in adjusting, higher and higher standards wee presented, and he was stimulated to meet them. The farmer and the social worker frequently shared experiences and ideas about him. These people became no only his ideals, but his conscience.

By the end of the second year he was able to do well, but this past reputation appeared to be against him. He was tried in a new cottage and told exactly why the change was made. Progress was then more rapid, the new cottage parents supplementing his two meaningful friendships.

The social worker then helped the boy reorient himself to his own past. By eliciting memories and piecing them together, the worker was able to help the boy understand that his misconduct was a consequence of his deprivations. He was told that while these explained his bad behaviour, society did not excuse it on that account.

He then became concerned about planning for his future. Farming was his vocational choice. In order to earn admission to a farm school he turned to academic schoolwork with a new interest. He went to a farm school after three years in the institutional an prepared himself there for a job.

He experienced some difficulties on a role but continued to maintain his friendship with the same social worker who helped stabilize him. He succeeded in being reasonably happy and self-supporting. On Dec. 9, 1941 at the age of twenty-one, he felt impelled to enlist in the United States Air Corps, three and one-half years after leaving the institution.

The Functional Psychotic

The functional psychotic deviates more widely from psychological normality than either the neurotic or psychopathic types, but he is also a product of a psychogenic etiology. Schizophrenia (or dementia paradox), the manic-depressive psychosis and paranoia, the major forms of functional psychosis, are found rather frequently among adult offenders. For the most part, however, they are disorders of maturity, less often found in the child and adolescent. In the first-named class of psychosis, however, and more especially in the hebephrenic form of schizophrenia, onset frequently occurs during youth. Hebephrenia usually begins before the age of twenty. Whether in the milder and nonpsychotic manifestations of the schizoid personality or in the more extreme break from reality that constitutes schizophrenia, these individuals tend to be socially unadjusted, withdrawn, and disorganized in their intellectual, emotional, and volitional functioning. They do poorly in school, at work, and generally in their personal relationships. Particularly, however, because of the lack of emotional stability, or integration, and their diminished self-control, these individuals may commit a wide variety of delinquencies. Often detectable in childhood or youth as increasingly disoriented prepsychotics, the excessively introverted schizoid should received early psychiatric treatment; with the progress of the disorder, any form of psychological therapy becomes increasingly difficult until, when he is completely disorganized, only medical help can possibly be effective) (mainly metrazol and insulin shock). Whether or not schizophrenia should in fact be considered psychogenic rather than constitutional in origin is a subject of dispute between the functional and neurological psychiatrists; there are strong evidences, at any rate, that organic changes do accompany the development of the disorder and some experts have attributed the pathology to hereditary origins. Whatever its roots, it constitutes a most serious disorder, not only psychologically, but in the delinquent and criminal manifestians that not uncommonly flow it. This case is illustrative:

A while, native-born American man, twenty-two years of age, single, attended school irregularly, lived a nomadic life, and worked only for short periods in a number of places.

As early as ten years of age, he was the source of a great deal of trouble both in his home and in the school. He rejected parental authority, resented the regulations of the school and was frequently in conflict with the principal, school teacher, and his schoolmates, he quite school at the age of fifteen or sixteen, but doesn't know what grade he had attained.

He had an uncontrollable temper and would become agreed at very slight causes. When agreed, he would leave home, sleep in the park, on the side walk, indoor ways, or go out in to the country. He would return in a and debilitated condition, sometimes after a few days and at other times dirty remaining away for two or more weeks. He was suspicious of his employer and fellow employees and frequently became involved in fistic s well as verbal combat. He would become dissatisfied ad quit or be discharged because of his uncooperativeness.

He had been in conflict with the law on tow occasions previous to the one preceding the psychiatric examination. On one occasions in an altercation he struck a man under the chink knocking him to the side walk, The patient insists the man died of fractured skull sustained from striking the side walk. The other appearance in court was the result of his driving to the adjoining state of Illinois in company with two other men in stolen car.

The offense for which he appeared in court previous to the psychiatric examination was attempted robbery in association with another man. He and his associate attempted to rob a filling station. The police department, having been advised of the possible robbery, had stationed officers in the filling station. When the would-be robbers had stationed officers in the filling station. When the would-be robbers came in, the officers opened fire. The patient's associate was killed, but the patient escaped unhurt. He was tried and sentenced to fifty years in the penitentiary despite the medical testimony, which was to the effect that he had dementia paradox and was limited in mental capacity to a level of high-grade feeble-mindedness.

Because of the customarily later onset, the manic-depressive affective or circular) and paranoiac insanities are less a problem with the juvenile than with the adult offender. But here, too, it is some times possible to diagnose the incipient manic-depressive in the school, the court, or the mental-hygiene clinic, in the moody adolescent whose affective and behavioural states display excessively wide variability. It is the opinion of the experts that the psychosis does not occur in children, but mature adolescents have been found in manic disturbance who are recalcitrant disorderly, rebellious, and aggressive. Similarly with the suspicious and ego-centric paranoid personality, the hazard of a more complete psychotic syndrome may be predicted from the incipient development of delusion a beliefs or the grandiose and persecutory trends a youth may manifest (Apparently no cases have been found of paranoid delusions in children.) Paranoid individuals, through they retain their intellectual powers intact, may be led to aggressive offenses against their imagined pesecutors. Juvenile delinquency is very rarely the product of a fully developed functional psychosis. However onset of mental illness does occur in youth-fairly often in the case of schizophrenia-and may be followed by laws violations resulting from the disturbed function. The following case illustrates the effects of a manic condition in a young man the exaltation and diminished self-criticism of this state commonly lead to this pattern of irregularities, extravagances, practical joking, and sometimes abusive and aggressive behaviour. A young man was charged with a breach of the peace by recklessly discharging a pistol at a ticket collector in a railway station. He had been in the habit of travelling to business every morning by a sub urban train. As a rule, the ticket collector did not ask him to show his season ticket, but on eh morning on which the offence was committed, the ticket was demanded, the patient refused to comply with the ticket collector's request, and when the latter insisted, the patient flew into a range and in a loud imperious voice demanded how the ticket collector dared ask a gentleman of his standing for his ticket when he knew quite well that he had a season ticket. When the ticket collector persisted in his demand for the ticket, the patient produced what after wards was discovered to be a toy pistol which made a loud explosion when it was forced, and discharged it in the direction of the railway official, who, thinking that his life was in danger, fled down the platform.

When examined by a psychiatrist, the patient was over talkative and exalted. He related with great pride and satisfaction the trick he had played upon the ticket collector, and also told of other practical jokes of a similar nature which he had recently carried out. For instance, on one occasion he told how he had thrown a bag of flour at a friend 9whoc was dressed up to go to a wedding) while they were both traveling in the same railway compartment.

The patient was clear in his mind. He had no delusions, and his memory and his intellectual functions were not in any way impaired. It was quite clear, however, that his behaviour was greatly influenced by his exalted mood; that his offence was the direct outcome of mental illness

The offender in the following case displayed paranoid trends inschizophrenia, which were primarily responsible for the murder he committed:

A young man twenty-two years of age was charged with murdering his sweetheart by striking her on the head with an iron bar. His history showed that there was no history of mental illness in his family. He was dully at school, and his teachers noticed that he was very timid and easily frightened. He worked for a time as an office boy, and later he obtained a post as an assistant store man in an engineering shop, which he held until his arrest. His parents had noticed that his behaviour had been peculiar for years. He had always been extremely as social and reticent. He spoke very little at home and when visitors came to the house, he either stayed in a corner by himself or he left the room. Before he left the house, he was in the habit of looking to make sure there was no one in the street. If he saw anyone standing near his house, he would turn back and wait until he could go out unobserved.

He had also been observed by his parents to stand in front of a mirror, partly undressed, grimacing, talking to himself, and masturbating. He had a violent explosive temper Some time before his arrest, in the course of an argument with his mother, he picked up the pocket and struck her on the leg. On another occasion his sister criticized the manner in which he was cutting bread, and the patient rushed at her with the break knife and attempted to stab her.

Another member of the family described the patient as "the kind of chap you never knew what he would do if her got into a temper." "His mind seemed to be miles away from the conversation."

A neighbour described how she had looked upon the patient as queer, and how she had seem him laughing to himself, and how at times when she met him in the street he would greet her cordially; at other times he would pass her without recognition. She had noticed that sometimes when he was walking along the street, he would take a curious little run. His employer thought that he was below the average intelligence. He had noticed that he was inclined to be dreamy, and that he seemed to have difficulty in grasping instructions. He was very moody, and given to violent outbursts of temper in which his language was obscene and threatening.

When examined, the patient was clear in his mind. His memory seemed unimpaired, and he was able to give an account of his history and of the events connected with his crime. He professed great affection for the girl he had killed, but he said that for some time he had been very disturbed in his mind because she would not agreed to marry him. She seemed fickle, and he felt that she had him "dangling on a string." It occurred to him that if he gave her "a fright" it would bring her to her senses, and so for some time before he killed her, he was in the habit of carrying an iron bar in his pocket with which he intended to threaten her when a suitable opportunity occurred.

On the night when the crime was committed he met the girl. He had the iron bar in his pocket. They ha a quarrel, and the girl said that she would not gout with him any more. She laughed at him in a scornful way. He struck her on the head with the iron bar which he had brought with him, and after that he said that he "lost his head completely." The patient's manner was very childish. He showed a complete lack of appreciation of the seriousness of his position. He said that when he saw any one bending down in the street, he had an impulse to strike him on the head. He also said that sometimes he go t out of bed in the middle of the night to look at himself in the mirror. He said that his own family regarded him as peculiar, and that his brother and sister sometimes called him "looney." While in prison awaiting trial, he developed further symptoms of mental disorder. He was observed to laugh and to talk to himself. He made peculiar gestures and grimances in a foolish way. He complained that the prisoners and the warders were talking about him and commenting about the shape of his nose, and he expressed the fear that he would be attacked. He said that he had written a song which he was going to submit to a composer for his opinion.

Emotional Disturbance and

Finally and the most important of the among the classes of psychological etiology that motivate delinquency, there are the psychological etiology that motivate delinquency, there are the affective responses of the individual to the thwarthing of basic needs in his personality. Emotional insecurity, feelings of inadequacy and inferiority, affectional undernourishment, extreme submissive or aggressive responses to rigid domination, these are among the common denominators not only of personal unadjustment but behavioral disorder and delinquency as well. The child needs an adequate and wholesome outlet for the expression of is psychic drives and wishes; as W.I. Thomas as pointed out, he requires particularly some satisfying implementation of his social drives or "wishes." Frustration of these basic psychological demands or the impossibility of securing socialized satisfactions may channelize them into law violative activity. Lack of affection and security easily lead to sexual delinquency, to theft, and incorrigible behaviour in children. Too, delinquencies provide an easy and often very full and exciting outlet for the boys' demands for status and adventure. If he lacks normal gratification in he home, school, and church, a child may discover in delinquency a fulfillment of his wishes beyond anything his moral elders can provide; and once a course of illegal conduct has been established in discharge of his psychic wants, their redirection into social sanctioned patterns is often very difficult.) The first case below illustrates the possible results of deprivation of security and affection:

Joe, aged fourteen, was referred to the juvenile court for stealing. He and a friend a year older had stolen guns, camping equipment, and an assortment of other articles, including a typewriter and fill in cards. The material had been taken over a period of months and had been stolen from various houses and stores, Joe, when questioned, explained that they had planned a hideaway in the woods for storage of the goods until they could be sold, Joe was obviously not the leader but he had been greatly intrigued with the details for their business venture. The typewriter had been taken from the school library so that all materials might be properly catalogued.

A study of the home disclosed that Joe was the older of two brothers. The father, an office worker, had shown little interest in him but had centred his whole affection upon the brother who was a year and a half younger than Joe and a quiet, confident child who felt secure in his position in the family. Joe was talkative, moody, and resentful by turns at home. At school he was smiling and popular, receiving high grades in his work. The mother had been a music teacher before her marriage and music was still her only great interest. One showed considerable ability along that line an his mother continually nagged him to practice on the piano. Both parents believed in strict discipline but disagreement in regard to methods led to constant bickering and inconsistent treatment of both boys. The father enforced his demands through threats and whipping while the mother scolded and entreated.

Joe, feeling insecure in the home and rebelling at the whole situation, readily accepted the plan of escape into the woods. On first contact with the probation officer he assumed a hostile attitude, reaction as he had towards his parents. When the probation officer ignored the challenge and asked him how he thought he could make the best of the difficult he was in, he appeared surprised, but he discussed his problems in a mature manner and offered constructive suggestions. He would work in the school library to pay for the material taken. He would spend his time reading in the evening, since the court restrictions would not allow him to go out. He did not like to be treated like a kid and if his parents would trust him with a gun he could go hunting with his uncle. He like music and he had been thinking about joining the glee club at school, but the wouldn't tell his mother so.

This study, from the cases of the Chicago institute for Juvenile research, reveals a delinquency ensuring largely from the desire for adventure and recognition:

When I was eight years old, I did my first job in the racket. This job was the biggest thrill I ever got in my life. It happened in April. That day I Was hanging around with the oldest brother and his gang. They had been playing baseball all afternoon and I was watching them.

When it got too dark to play ball well went into the alley to have a smoke and tell stories. The big guys got to talking about stealing, and my brother said he had good place spotted where he could get some easy "dought." The place was a butcher shop in thirty-first Street. The big guys planned everything, and I only listened. These guys were seven or eight years older than me and had pulled off a lot of big jobs before. They would never let me go with them on big jobs; but this might I went along and they didn't say a word We all when to the butcher shop about 11.30 o'clock. It was very dark and everything was quiet, and I was nervous and stayed close to my brother and another big guy went up to the building to see if the doors were unlocked. My brother had been in the place a few days before to see how to get in and where the cash register was; and so he led the way. I and two other guys waited close to the alley between two buildings. We were going to give "jiggers."

In a little white my bother came back and said everything was locked tight. The owner live over the butcher shop so we couldn't make much noise by breaking the glass or jimmy the door. We all went up to the back door, and then my brother got a box and stood on it and tried the transom and it opened. It was too little for my brother or the other guys to get through. Then I was thrilled when they said I'd have to crwawl through the transom. That was the kick of my whole life.

I was only eight and always was very little, so I could get through the transom easy. I was scared but made up my mind to go through anyway. I was too thrilled to say no. My brother lifted me up on his shoulders and I crawled through the transom. I hung down on the inside and stood on an icebox and then crawled down on the floor. The door was looked with a padlock and chain, but I was able to unlock the window and let the bi guys in that way. The big guys looked for money first and found $22. Then we all got everything we wanted to eat and several cartons of cigarettes and ditched the place.

When we got out, my brother divide up everything and I got $4 and a lot of cigarettes. I felt like a "Big shot" after that night and the big guys said could go with them every time they went Robbin." Almost every night we went robin." And many times I had to craw through transoms and one time through an icebox hole. That's hole. That' why they big guys called me he "baby bandit." The significance of the role of emotional discomforts, of the frustration of basic psychological needs, which so commonly reflects affective inadequacies in the child's home, has been given all too little attention in etiological research because of the preoccupation with the objective factors in the environment and the more behaviours traits of personality. John Dollard, However, has attributed a primary importance to the mechanism of frustration and consequent aggressive reaction s against the environment." The pattering emphasized in his work is a crucial determinant of delinquency in many cases, but this explanation is too pat in its simplicity for universal applicability.

Healy and Bronner's more analytical investigation of the diverse chancelling of emotional requirements that may be thwarted in the family and of the valid reactive mechanisms that reflect the psychic-constitutional differences among children responding to their equal various situational milieux provides a more adequate appraisal. This research is considered in some detail below, with its implications for the psychological and social influences upon misconduct.

Intrafamiliar Tensions

In their work New Light on Delinquency and Its Treatment, William Healy and Augusta Bronner have expounded a theory of delinquent behaviour based upon psychological, psychiatric, and casework investigation. Although the sample of offenders selected for study was small (153), it nevertheless constituted an excellent group for the purposes of their study, since it represented delinquents in three cities (Boston, New Haven, and Detroit), studied clinically by three teams of experts (psychiatrists, psychologists, and psychiatric case workers) who pursued their analyses far more intensively and extensively than is customary in etiological research. The cases selected all had non-alignments siblings to whom comparison could be made as a control group with social back grounds similar to those of the offender sample. So selected, these cases did not make up a representative sample of delinquents. This limitation was not of primary importance, however, since the main purpose of the study was not o describe a typical group of offenders, but to analyze in depth the conduct motives and resultant in delinquents and to discover the differences marking their nondelinquent brothers whose general situational patterns in the home were similar. Thus it was hoped that more light might be thrown on the significant, but in the main obscure, problem, why two or more children in the same family, exposed to roughly similar influences, should behave very differently. The study has the additional virtue of putting its analytical findings into numerical terms for comparison with the nondelinquent brothers and with other samples of offenders. Thus combining the virtues of careful ongitudinal and cross-sectional research and permitting subsequent check by ofter investigators on the accuracy of their finings and the validity of their conclusions.

To limited extent the observations made in this research overlap the analyses of other students who have found causation in both individual and social factors. This study, however, highlights, as does the procedure of the psychiatric clinic itself, the interrelationship between the two categories of influence, their inseparability in the process of causation. Thus the study finds, and attributes an important significance to the deleterious social factors that appeared in man of the cases: In only twenty-two instances of their 153 delinquents did they find the child living in a favourable situation as to the following:

1. Reasonably good home conditions from the standpoint of stability, living conditions, normal recreational opportunities.
2. Reasonably good family attitudes from the standpoint of avoidance of friction, rational treatment of children, and being law abiding
3. Neighborhood not distinctly bad from the standpoint of inimical influences.

In only nineteen additional cases were two of thee categories of influences favourably. They found evidence of the importance of these situational variables in the fact that there were significantly fewer cases (a proportion of one to nine) of multiple delinquency in the family, when the family and neighbourhood conditions were favourable.

The workers also investigated the a subjective personality factors in the sample an offered striking data on developmental history, psychological deviations, and physical habits in contrasts drawn between the delinquents and the controls. These gave rater clear evidence of the relevance of constitutional and psychological factors in the making of the delinquent, revealing the much higher frequency of defect and disorder among the delinquents.

These findings as to social and individual circumstances add confirmation to the discoveries by other scholars of frequently recurring elements in delinquent histories. However, this study proceeds further to focus attention upon the emotional adjustments of the offenders, the quality of their subjective responses to the external milieu. Considerably more striking than the data above was their observation that 91 per cent of the delinquents suffered from major emotional disfunbances, 75 per cent displayed a marked antipathy to parents fro to school, whereas only 13 per cent of the no delinquents experienced any form profound emotional distress. The main forms of maladapted affective responses among 143 delinquents accepted as cases for treatment were these.

- Deep feigns of being thwared in self-expression and other self-satisfactions—45 cases
- Feelings of being rejected, unloved, or insecure in affectional relationship -53 cases
- Marked feelings of inadequacy r inferiority in some situations or activities—62 cases
- Emotional disturbances abut family disharmonioes, discipline, etc.—43 cases
- Great persisting sibling jealousy or rivalry—43 cases
- Deep-set internal emotional conflicts-19 cases

An unconscious sense of guilt and feeling of need for punishment-discovered in 11 cases.

Healy and Bronner conclude, after detailed analysis and comparison of their cases and controls, that the delinquents, more often thwarted, rejected, and disturbed, find substitutive satisfaction for their psychological needs in channels that lead to delinquency. The nondelinquents, less often from vicious homes and rarely disturbed emotionally, express their less distorted needs in socially acceptable ways, without excessive or escape techniques. Their greater docility and lesser gregariusness, as revealed in the study, facilitated outlets and reactive manifestations of their needs that are socially defined as moral and legal. Evidence for the soundness of this theory that delinquency is a channel from is directed drives is presented in abbreviate case history comparisons. The following is enlightening particularly as to the combination of constitutional and emotional influences.

There were eight pairs of nonidentical twin in the Healy and Bronner sample. IT was hoped that through this group they might uncover factors that would permit better prognostication of delinquency. Conceivably, what with the similarities between the twins, the differences significant for delinquency might be segregated. Here again, however, they could find no clearer composite for prediction than the combination of abnormal personality with deeply felt thwartings and inimical family situations. These twin studies also pointed up the close relationship between emotional disturbance and the onset or continuation of delinquency. This case is illustrative:

We may begin with our six-year-old twins. Although so young the delinquent had already definitely run away twice and on numerous occasions had wandered from home, having been picked up by the police several times. On one occasion he made a serious attempt to set the house on fire. He was a truant from school and had repeatedly stolen, even considerable amounts, from teachers as wells from his parents. He had also stolen money from a neighbour's apartment, a hospital, and a gasoline station Moreover, his parents complained of his disobedience, impudence to his mother, and his temper tantrums. In general, he was regarded as being extremely trouble some at school and incorrigible home. The control was an eve-tempered, obedient, thoroughly well-behaved 'model child."

The twins in this family of three children were the result of a welcomed pregnancy; the control was born first and without instruments; the delinquent was delivered by instruments during a difficult labour. They weighed about six pounds each; they were both bottle-fed from the time o birth. The delinquent was a fussy crying baby; the control was not. They both had an uneventful health history and were of about equal size and in good physical condition at the time of our study; they both exhibited occasional enuresis, but neither had any nervous habits. Our psychologists gave the delinquent an intelligence rating of I.Q. 82, commenting that probably he would have done considerably better if his attention could have been fully held. The I.Q. of the control was 95. The summing up of the personality characteristics of the delinquent shows that was exceedingly restless, active, impulsive, uninhibited, continually running off to play with other boys, among whom he was regarded as a follower. He frequently acted as a "show off," being specially delighted with the attention he received because he was such a small boy when he was held in the detention home for a new days, and he was especially proud of his grinning picture in the newspapers, which had appeared once after he bad been found by the police. At home he was said to be extraordinarily inquisitive, "into every thing". In the physical examination he appeared to have some shame about his body.

The control was an amenable, pleasant youngster, much quieter, self-reliant, and easily amused at home, preferring to be with his mother rather than to play with other boys, although he did sometimes gout to play.

The family lived in crowded quarters, but there was no

great financial stress. Earlier they had lived in better homes and in decent neighbourhoods, though they moved much because the father frequently sold the home. The parents were thoroughly in compatible, engaged in much open quarreling and often, been before the children, made counter accusations and threatened divorce. There was violent disagreement about discipline; the mother charged her high tempered husband with whipping too severely and he maintained that although she was rather well educated, she was really dull, incompetent, and too lenient.

An amazing difference of attitude on the part of the father towards the delinquent and the control was reported to us and acknowledged by him. He maintained that the cause of it was the stubborn lack of response to him shown by the delinquent when an infant, only two months old. He said he detested they boy then and had ever since—"I can't bear to have him touch me. I would rather have a snake around me than have him." Though the mother stated that the father always repulsed the boy when the latter attempted to climb on his knee, both parents insisted that the child had never given expression to any strong feelings about being is liked by his father.

The older boy of eight had been very jealous of the twins from the time of their birth and cruel to them. He suffered from epileptic attacks for two years and afterward simulated attacks in order to gain attention. Though particularly hating our delinquent, or perhaps because of this hatred, he entered into delinquencies with him and inducted him into sex practices. Asia result of frequent beatings and the family disturbances, this older boy had emotional tensions of his own and, besides, gave evidences of being an epileptic personality.

Both the delinquent and the control had plenty of toys and at times had a playroom to themselves. They both went to Sunday school fairly regularly and willingly. The delinquent freely unburdened himself to the psychiatrist, telling many of his delinquencies, some of which were not known. About stealing a considerable sum from his teacher, he said rather demurely that his dad had to pay it back. In childish fashion he told of "other kids" inciting him to set the house on fire. His story included "another boy" with a name Similar to his own- "a bad boy" "He was mad at my dad and set the house on fire. Then I was afraid, and we got the first mostly out b stepping on the curtains before the firemen came." He went on to give in the plainest language the vile epithets which his father called him and also the equally opprobrious terms that his parents applied to each other. He showed much feeling about all this, saying that he was going to hit his father "right on the chin" some day. He also evidenced great dislike of his older brother and shame about their sex play together. The control, he said never did such things. It came out very clearly that the boy felt extremely insecure in his family relationships and that he greatly created recognition and affection.

The control, on the contrary, was very sure of himself and quite well satisfied with things as they were. In speaking of his brother to the psychologist he priggishly boasted, "I am the good boy."

New Light on Delinquency and Its Treatment has been criticized on the ground that "emotional disturbances" are too vaguely defined and too easy to discover in or project upon a sample of individuals known to be offenders. Here one encounters the familiar and apparently inescapable problem discussed in Chap. III: in any investigation of cause, whether case history (or other longitudinal method) or statistical (cross-sectional), one must rely on the observations of imperfect investigators. It is true, more cover, that this study focuses attention on subtile variables of feeling and relationship that do not lend themselves to easy or exact measurement. While ship that do not lend themselves to easy or exact measurement. While it must be recognized that clinicians do operate with such intangibles an undoubtedly with certain biases implicit in their frames of reference, yet this study must be accredited with an unusual amount of case in the analysis and comparison of the delinquents an their controls. Here the expertly trained investigators were looking for no single factor or set of factors; they sought widely, intensively and empirically, it appears, for any deficiencies and advantages that might be discovered as significant in distinguishing the etiologies of delinquent and nondelinquent. Though the size and selection of their sample limitstatistical validity, the study was primarily concerned with the dynamics of the individual offender, not description of the mass. Further studies of this sort, costly as they are, and badly needed to extend knowledge of causal processes. But the conducting of this study in more than one community with a number of experts functioning independently in the investigation lends additional weight to the Healy and Bronner conclusion, it is improbably that their observations could be the product of similar consistently erroneous diagnoses by a series of trained psychiatrists, clinical psychologists, and psychiatric social workers. This provides a check of internal consistency, although the authors did not exploit it in proof of their findings.

6

EDUCATIONAL SIGNIFICANCE OF EMOTIONAL DEVELOPMENT

We are all emotional beings – more emotional than we often realise. Most of the affairs of our everyday life are tinged with feelings and emotions. In the home or the school, on the playground or among associates, emotions are aroused. They play a dominant part in determining our actions and reactions. Life without emotions would be dull and dreary. Emotions add colour and spice to living. If there were no joys and sorrows, excitements and disappointments, much of human life and experience would have lost their colour and warmth. One cannot conceive of what children's lives would have been without emotions. When the teacher enters the class, he finds his pupils in different emotional states. Some are happy because they received birthday presents. Some are worried because they failed to complete the home task. A short while ago two of the children quarreled with each other and are angry at each other. The teacher himself is in a particular emotional state pleasant or unpleasant. Emotions are the dynamics of behaviour.

Educators should carefully deal with them. Whether it is solving a problem in mathematics, map-work in geography or appreciation of poetry – the teaching – learning situation always gets emotionalised. From home and neighbourhood, children bring with them different emotional states to school.

These are increased, modified or calmed by the emotional tone of the school and the class. Hence the teachers have to set such an emotional tone that it would help the development, learning and adjustment of children. They have to build up a secure, pleasant and constructive atmosphere both in the class and in the school for this purpose.

Emotional qualities of the teacher and his teaching affect the mood of the pupils and the atmosphere of the learning situation. The teachers also bring with them to school emotions that colour their personal lives. These emotional states partly determine the atmosphere they build in the school. If teachers with wholesome adjustment in life situations come to the school with emotional equipoise, the atmosphere of the school gets elevated and if they themselves are emotionally imbalanced or disturbed to a considerable degree, the emotional tone of the school degenerates.

To give a concise definition of "emotion" is too difficult. It is many things at once. When we speak of "emotion", we usually refer to:

(i) the subjective feeling, that is, the way we feel when we are emotional;

(ii) the bodily states, that is, the physiological changes that take place when we are in different emotional states;

(iii) the mode of expression; that is, how emotions are expressed by language, gestures and facial expressions; and

(iv) the motivating force, that is, how drives, motives and emotions are very closely related concepts and determinants of all our actions and reactions.

Emotion is a conscious experience. Anger, fear and joy are conscious experiences. It is an excited state of the individual arising in response to a stimulus the object of emotion. Some are of the opinion that an emotion is an excited state arising in response to stimuli for which the individual has no adequate readymade reactions. The stimulus sets off energy mobilisation in the individual. The energy is channeled into movement towards or away from the object of emotion, taking it or leaving it. Energy mobilisation begins when the inner equilibrium is disturbed and continues until the equilibrium is restored. In psychology, this is called "*hameostatis.*" Objects that arouse "*Positive Feelings*" like delight and joy restore equilibrium. Stimuli which "*Induce*" fear or anger threaten the loss of equilibrium. The emotional experiences can be expressed in "*introspective reports*". Let us consider an example.

On his way to the college, an adolescent boy turns a bend along the road and finds a big bull standing. Suddenly, the brute comes charging at him. The boy perceives the situation but does not have any adequate readymade reaction for the charging bull. He gets excited and physiological disturbances take place in him. He feels a disturbance in the viscera, his blood circulates very rapidly and his heart pounds and body trembles and sweats profusely. The need to avoid pain and injury generates in him energy that gives a direction and purpose to his behaviour. He must escape the threatening situation at any cost. Suddenly he turns around and begins to run. While running for safety he meets a friend who enquires about his panic, halting him. His speech fails him and he fumbles and falters but drags his friend along with him for some distance more. When he and his friend are relatively safe from the charging bull, the boy begins to calm down and starts telling his friend about his emotional experience. *The Emotion of Fear.*

Classification of Emotions

Thompson (1965, p. 274) classifies emotions into two types: (a) The *pleasant-integrative* and (b) The *socially – disintegrative*. He defines the former as: "psychophysiological states which tend to accompany moderate muscular tension and verbal reports of pleasure'. We tend to seek and maintain this psychophysiological state of affairs both for the moment as well as a life-goal. Elation, affection and hope are pleasant, positive and integrative in nature.

Thompson defines the latter as: "psychophysiological states which tend to accompany extreme muscular tension, heightened smooth musculature response and verbal reports of displeasure. We are averse to these psychophysiological conditions and tend to avoid or abolish them whenever possible. The strong emotions of fear, anger, jealousy and disgust fall into this category.

Theories of Emotion

A detailed discussion of different Theories of Emotion is beyond the scope of this text. Yet to treat them as briefly as possible:

The James–Lange Theory, originated by William James and Carl Lange maintains that the emotions we feel are the result of our perceptions of the changes taking place in the body during emotion. Thus, William James writes, "The more rational statement is that we feel sorry because we cry; angry because we strike; afraid because we tremble and not that we cry, strike or tremble because we are sorry, angry or fearful as the case may be". The Cannon-Bard Theory of emotion, originated by Walter Cannon and Philip Bard says that the felt emotion and the bodily reactions are independent of each other; both are triggered simultaneously and are both controlled by activity in the central nervous system. These two theories seem to focus on the relationship between felt emotions and bodily states.

"Bodily sensations, particularly the activity of the autonomic nervous system, are important to feelings of emotion, but different emotions show pretty much the same physiological patterns...The individuals' appraisal of the emotion-producing situation is an important factor in determining the emotional response" (Hilgard et al., 1976, p. 351). Therefore, S. Schachter and J.E. Singer propose that an emotion we feel is our interpretation of an aroused, or stirred up bodily state. Emotional states are a function of the interaction of cognitive factors and a state of physiological arousal. The Schachter. Singer theory of emotion may also be labelled as Cognitive physiological theory of emotion.

R. Plutchik's theory of relationships among emotions tells us that there are certain primary emotions which are derived from evolutionary processes. Therefore, they have adaptive value. These primary emotions can be mixed together and arranged spatially to show their intensities their similarities and their differences.

Proposing one theory of emotions and motivation R.W. Leeper says that emotions are motives which give direction to behaviour and keep behaviour going. Affective processes are pleasant or unpleasant, Emotional processes along with other motives have important perceptual aspects. S.S. Tompkins maintains that emotions provide the energy for motives. Drives and motives give information about some need or condition of the body. Emotions or effects accompany these drives and provide energy to the latter. Emotions amplify drives to give them their energy. Distress amplifies hunger to give it its strong motivational power.

Richard Lazarus and his co-workers forward a cognitive theory of emotion. "The theory says that emotion is a complex and patterned physiological responses and a set of tendencies to action that are triggered by a person's appraisal of events in the environment...the cognitive theory postulates that distinct physiological patterns exist for the different emotions it also assumes that generalized arousal occurs in emotional states it claims that the emotions a person feels comes from appraisal of the external situation, appraisal of the physiological conditions, and appraisal of the consequences of the action that might result from the emotional state."

Growth Forwards Emotional Maturity

Emotional growth is dependent upon maturation. General excitement is the first pattern of emotional expression to appear in the infant. As the infant grows in age, from the parent trunk of general excitement, there is an increasing differentiation of emotions. The physiological disturbances that accompany an emotion remaining the same, with increasing age in children, changes occur in the situations that arouse emotions and the mode of expression of emotional reactions.

With increasing age, there is a decreasing number of situations and objects arousing emotional reactions, particularly in so far as the emotions of anger, fear and affection are concerned. This happens on account of the children understanding better and to a greater degree, objects and situations, as they grow older and older. The child fears thunder and lightning at first but later when he understands the cause, this natural phenomenon does not arouse the emotion of fear in him any longer.

There are three reasons for reduction in number of situations and objects arousing the emotion of anger in children with increasing age. In the first place, with growing ability children learn to manipulate material objects, learn the rules and techniques of playing games and learn how to get on with their seniors and leaders. The second reason is that as their understanding grows more and more, they appreciate the motives of others and avoid hurting the feelings of the latter. In the third place, greater and greater realisation in them as to what things they can do and what they can not; makes them not to set goals that they cannot attain.

Objects arousing the emotion of love and affection change in kind and number with increasing emotional maturity of the child. The infant develops love and affection towards his own sell at first. This is called Self-love. Them the parents become the love objects of the infant. But in course of time, the child wants to be emotionally independent of the parents. A time comes when his affection for other children of the same age and sex develops. This too changes. When he grows into late adolescence, he begins to concentrate upon a member of the opposite sex as his love-object. Adolescents who seek emotional dependence upon their parents are treated as emotionally immature.

As he grows towards emotional maturity, emotional responses of the child go on changing. At first, his emotional responses are rather crude but as time passes on they become more and more refined. For example, if you block the play-activity in progress of the child, he may come running towards you, open mouthed, ready to bite or may cry and roll on the ground. On the other hand, if you offend an adolescent thwarting any activity he is engaged in, he would argue with you or frown at you rather than harm you physically. As they grow more and more mature emotionally, children take recourse to diplomacy in setting their disputes.

While children give immediate expression to their emotions, individuals with emotional maturity learn to control their emotional responses. When angry, children may kick, hit or bite, When afraid, they may cry, freeze or seek protection from the mother; and when happy, they may launch, dance or clap their hands. In course of time as they find that adults do not appreciate some of those responses, they learn to substitute other responses for the former. Those who are emotionally mature, exert control over their emotions. They may frown when they are angry, expressionless when afraid and simply smile when happy. Some others may remain expressionless even if they are angry, happy or afraid.

An individual who is emotionally mature, has learned how to meet and sole his own problems. He faces reality as it comes to him. He does not try to escape from the problem situation but resort to defence mechanisms such as day dreaming or rationalisation. A child who has not prepared himself to take an examination may avoid it and subsequently declare that sickness disallowed him to sit out the examination. On the other hand, an emotionally mature person would try to perform a task whether he succeeds or fails in accomplishing it.

A child whose emotional needs are satisfied in desirable ways, developes into an emotionally mature member of the society. If his emotional needs do not get satisfied in acceptable ways, a child may develop abnormalities of behaviour; and in extreme cases there may be disintegration of his personality factors turning him antisocial at last.

The school and the teacher should help the child satisfy his emotional needs properly, and live a life to the best satisfaction of all. This shall make him healthy and mature emotionally and finally make him a useful citizen.

Sequence of Emotional Development

"Perhaps the nearest approach to a useful series of developmental stages for the first two years of life comes from an investigation by Bridges (1932)," states Stott (1978), p. 281). One of the most important outcomes of the study was, "emotional responses gradually become more differentiated."

In relation to '*differentiation*'. Bridges described with approximate timing, the sequence that takes place from birth to 2 years of age. According to this study, the first pattern of emotional expression that certain kinds of stimulation give rise to in the infant is a *generalised overall excitement*. Bridges regarded this as the original emotion.

The general excitement pattern does not persist too long. While the infant is about 3 weeks old, the first step in differentiation takes place. Distress or the unpleasurable excitement stems out of the parent trunk of general excitement; and by the time the infant is about 3 months old, delight, the distinctly pleasurable sort of excitement becomes differentiated. Delight is characterised by relaxation, smiling and cooing in response to some satisfying situation whereas distress is marked by muscular tension, crying and checked breathing in response to certain disturbing stimuli.

The next differentiations in, the sequence are on the unpleasurable side. According to Bridges when the baby is about 5 to 6 months old, distress gets further differentiated into anger, disgust and fear.

At about 7 months, elation becomes distinguishable from generalised delight; and by the eleventh month there are clear signs of affection for adults. By age 15 months, the baby makes an affectionate response to other children as well as to adults. The infant expresses elation in response to objects and events and affection with regard to persons.

When the infant is about 18 months old, jealousy appears as a further differentiation of distress. Finally, at about 20 to 21 months, the pattern described as joy makes its appearance.

In another study Bridges speculates that by five years of age the emotions of shame, anxiety, disappointment and envy are differentiated from the distressful stem, while the emotions of hope and parental affection come out of the parent stem of delight.

Various emotional expressions found in adult life are acquired through further maturation and learning in the subsequent childhood years and during adolescence. Emotional responses change as a result of ageing. For example, anxiety may change into grief, fear into worry and affection into benevolence.

In infancy the emotional tensions and satisfactions of

the child are closely tied to his organic needs and dependence upon the grown ups. The infant needs his hunger satisfied, wants to be kept warm and dry and have as much sleep as he requires. He needs sucking and elimination of body waste. He needs to have muscular activity and locomotion without restriction. These include kicking, thrashing of arms, creeping, walking, running and jumping. He wants to discover different parts of his body and is curious about them.

The infant depends upon his mother, nurse and other adult members of his family emotionally. From them he demands both physical and verbal attention. He needs to give affection and get affection. He should feel that he has a definite place in the family and is secure. He wants to perform tasks for which he is ready and expects adult approbation for his accomplishments. These are some of the most important needs of children demanding satisfaction. Emotional tensions arise if they are thwarted.

"The emotions most common and certainly the most revealing are anger, fear and love", writes Jordan (1959, p. 93). Let us take love first as it is the most powerful emotion in the life of an individual.

The Freudian Stages of Psychosexual Development

"The series of stages in the psychosexual development of the individual adopted by many psychologists as the basic framework of personality development are the *oral*, the *anal*, the *phallic*, the *latency*, and the *genital* periods....The Freudian sequence of phases in psycho-sexual development, as we have noted, has mainly to do with change in sources of gratification and objects of affectional attachment. For many students of personality and particularly for those engaged in clinical diagnosis and treatment, this conceptual sequence has been found useful and meaningful, and has been widely accepted," says Stott (1978, pp. 277-281). But things do not stop here with this.

Psychosexuality involves the development and broad biological, social and psychological manifestations of the sex impulses. It forms the nucleus of psychoanalytic theory and treatment. Psychologists recognise three principal stages in the sexual growth of the individual. The first is the period of *Infantile Sexuality,* extending from birth to about age six. The second stage is the *Latent* period, from age six to twelve. The third, the *Genital Stage,* starts with puberty and extends through adolescence to maturity.

(1) *Infantile Sexuality* which lasts from birth to age six has three stages—the *Oral,* the *Anal* and the *Phallic.* The Libido (energy of the sex drive) is fixated in these three *erogenous Zones* of the body during this period. The erogenous Zones are those parts of the body that can be stimulated to respond erotically, especially, the oral, anal, and genital orifices (outer openings) and organs where the sensitive mucous membrane comes close to body surface.

(a) The Oral Stage

In the oral stage, approximately the first two years of life, the mouth and lips are the major sources of pleasurable sensation. The taste, tactile and kinesthetic sensations accompanying sucking, swallowing, biting and eating provide a primitive kind of erotic gratification. This stage is also called the autoerotic period. The functioning of the libido mainly revolves around the physical self. The Pleasure Principle is overwhelmingly dominant.

(b) The Anal Stage

This stage, the ages one to four, involves pleasurable tactile and kinesthetic sensations in the anal erogenous Zone. An infant does not react with an adult's disgust to defecation. He is messy, likes to smear and enjoys anal stimulation. Pleasure is derived from expulsion and retention. As the infant is chiefly concerned with the satisfactions associated with his toilet habits during this period, Freud believed that these characteristics pointed to the existence of an innate anal sexuality. Between the ages two and four, the infant becomes conscious of himself as an independent individual and directs his libido upon himself as a psychological entity. This self love is termed *narcissism* the Pleasure Principle is beginning to be controlled by the growing Reality Principle.

(c) The Phallic Stage

This stage, from about four to six years, is marked by the dominance of genital sensations, as the libido is localised in the genital organs. Infantile masturbation is not uncommon. There is the development of Oedipal involvements with parents. This means that the libido object shifts from the self to the parent of the opposite sex. Little boys fall in love with their mothers and dislike their fathers. Little girls love their fathers and hate their mothers. This triangular "family romance" – mother, father and boy or girl is the well-known *Oedipal situation* or *Oedipus complex*. In the girl, it is sometimes referred to as *Elektra Complex* or *Elektra situation*; often, however, the term Oedipus is used for both sexes. This peculiar parent relationship is soon dissolved in normal children and they relinquish all infantile sexuality. The Reality Principle is now well established.

(2). *The Latent State.* The Latency Period lasts from ages six through twelve. During this period the typical child becomes concerned with school, athletic, or other achievements. He renounces his earlier sex interests. Emphasis is now placed on intellectual growth and social and moral development. He forms friendships with peers of the same sex and avoids the opposite sex. This is an age of high ideals, altruism and prudishness. He puts emphasis on "the rules of the game" and "not to step on the cracks in the pavement". There is a decline in narcissism. He resents excessive parental display of affection though he is fond of both parents.

(3) *The Genital Stage.* This stage starts with puberty and extends through Adolescent to maturity. With the arrival of puberty, the first phase of adolescence, latency ends and the genital Zone attains permanent primacy over the other Zones.

(a) Revival of Infantite Sexuality

However, the onset of puberty in the early teens brings about a revival of the oral, anal, and phallic stages. Early adolescence is also a period of self-love. Adolescent boys and girls secretly admire and are interested in the body. Thus, the onset of puberty revives auto-eroticism and narcissism.

(b) Homoeroticism

The period between twelve and fifteen years is the *homosexual phase.* It is often marked by attachments to playmates and friends of the same sex. The stage of 'Pals' begins in late childhood and continues through early adolescence. Girls develop 'crushes' on each other or physical contact with their teachers. They send flowers to one another as a mark of friendship. Boys form gangs, have special pals and heroes. Youngsters at this stage are very egocentric and exhibitionistic. In the pregenital stages the individual is narcisstic for gratification is obtained only from stimulation of his own body, and other persons are valued merely because they heighten his own pleasure. The fondness of the youngsters for like-sexed individuals and fear of the opposite sex is called *homoeroticism.*

(c) Heterosexuality

In the heterosexual phase, the object of love is another individual of the opposite sex. Early adolescent love affairs are largely narcisstic and experimental in nature. Boys and girls demonstrate their ability to attract the attention of the members of the opposite sex. At the same time they gain confidence in their sexual maturity. Therefore, it is not at all surprising to find them exhibitionistic to a considerable degree. In the genital stage narcissism is replaced by altruistic identifications with people who now come to be valued for their own sakes. Serious love attachments involving a heterosexual exchange of libido appear in the late teens and early twenties. There is a marked revival of the Pleasure Principle in the early phase of the genital period, but eventually the Reality Principle regains the upper hand. "Pregenital impulses remain alive in the mature adult but they are subservient to and fused with genitality; Each stage thus continues as an undercurrent in succeeding ones; the Freudian developmental eras are not sharply separated but have residual, permanent effects," state Rosen and Gregory (1965, p. 63).

The child has to move through all the stags of psychosexual development naturally and easily so that he shall have a wholesome emotional development. When there is any kind of arrest on any stage, deviations in normal behaviour may appear in later life. As was stated earlier, psychosexual theory has far-reaching significance for the understanding of human behaviour. Freud believed that many character and personality traits of normal individuals are derivations and manifestations of infantile sexuality.

Anger is one of the emotions experienced by children and adults alike. Anger includes emotional states ranging from the milder form of resentment to range which is an extreme form. Thwarting of, objections to or interference with ordinary routine habits such as those connected with their feeding and toilet activities may cause in the infants and younger children emotional outbursts of anger. Difficulties in social relationships, direct conflict with authority, forcible restraint blocking activities in progress, thwarting of interests, wishes and plans and purposes arouse anger in the infants and young children. The baby gets angered if he is forced to go to bed while at play. He shall be too angry if he is stopped playing with the mechanical toy.

When he is angry, the infant screams, kicks stamps, and throws his body about. These responses are direct and random. As he grows older, the child discovers that these responses are frowned upon and resisted. He discovers that it does not pay to kick, bite or stiffen his body. Therefore, in course of time he *learns* to employ subtle techniques such as pouting, dignified withdrawal or use of caustic language in expressing his anger.

When he is about six years old, the child enters the elementary school. This is a new period of emotional development for him. His activities and interests differ much from that of the infant. Though all the six-year olds enter into the elementary school during the same period, they differ widely in emotional maturity and behaviour pattern. From age 6 through twelve, the child emphasizes on intellectual, social and moral development.

The situations arousing anger in the child decrease in number. However if you force a child playing a game of football he would certainly be angry and might express his anger in less aggressively physical ways. Perhaps, he would say, "I don't like you" or "I hate you".

Adolescence is a period of transition from childhood to adulthood. Therefore the status of an adolescent is very ambiguous. This makes him involved in emotional conflicts.

Any social situation that places the adolescents in an unfavourable position arouses anger in them. They are angered if they feel that they are insulted in public. If the teacher is critical of any adolescent boy or girl, that places him or her in an unfavourable position in the class causing embarrassment, that would arouse anger in him or her. Ignoring him if the teacher appreciates the achievement or ability of others, the adolescent boy becomes angry. If some one unfavourably comments on her physical appearance, an adolescent girl gets angry. If their clothes or other possessions are spoiled by chance, adolescents are much angered.

When they are angry, emotional responses of adolescents are different from those of early childhood. While two young children fight with each other when they are angry, two adolescents engage themselves in arguments or fuss about or pass uncomplementary remarks on each other to express their emotion. But occasionally, two adolescent boys lose their temper, swear, and fight, Adolescent girls stamp their fect, scream and scratch each other. Anger in adolescents may last from one minute to 48 hours.

Most of the things that anger the adolescents also anger the adults. Impositions upon the, interference with their personal belongings or habits cause anger in adults. When injustice is done to them or when others are given the credit for the work they actually did, adults express their anger. Very often they express a controlled emotional response, such as a sarcastic remark or a sharp criticism or simply remaining silent with a frown. However, occasionally they fly into a range and do not shrink to commit acts of violence, which may range from assault to murder.

Fear is another important emotion that children very often express. Fear is closely related to inherited structure as well as the process of conditioning and unconditioning in courses of different experiences of the child.

A variety of objects and situations produce fear in infants. Strange animals, objects, persons, situations and noises cause fear in infants. Also falling objects, persons and objects associated with painful experiences, sudden and unexpected movements, the danger of falling and bodily injury produce fear in infants. They are also afraid of the dark, being alone and of imaginary creatures and very loud noises.

The typical response during the emotion of fear is freezing or flight. In the early years of life, infants cry, tremble and blanch when afraid. Very often they avoid the fear producing situation; but with increasing age, he infant instead of avoiding a fear producing object may try to assault it and while experiencing a fear producing situation may verbally report: "I am afraid.".

During the elementary school years children are afraid of the imaginary creatures. They are afraid when they have fearful dreams. The fear that they might meet with accidents, that they might lose their objects or that robbers might raid their home. They fear that their parents might fall ill, that they might fail in the class examinations and school examinations and they might be insulted in the class. Born out of sheer imagination, these fears and apprehensions worry them much.

Social and personal matters cause fear in adolescents. They are anxious about their physical appearance, and the impressions they make on the boys and girls of their acquaintance. Thus, inadequacies in social and physical situations, real or imaginary, cause fear in adolescents. Another cause of fear during adolescence is related to sex matters. The late maturing boys and girls apprehend that they may not mature sexually at all and they are afraid of it. Sexually mature adolescents occasionally satisfy their sex need by unnatural means and later fear that it would cause bodily harm in them. This conflict in them between the urgent satisfaction of sex need and the resulting physical injury often leads them to unwholesome adjustment. Adolescents fear the loss of prestige in social situations as well as the uncertainities of future occupational situations. Besides these, the possibility of their failing the examinations or meeting with fatal accidents or getting murdered or poisoned or not getting sufficient money to spend are some of the other sources producing fear in adolescents.

Educational Significance of Emotional Development

Parents and teachers should respect the emotions of

children for what they are. Anger, fear and love are very powerful emotions that motivate the behaviour of all the individuals. They should be channalised in the proper direction for development of an integrated personality and wholesome adjustment.

Anger, fear and love have significant values. Fear makes an individual cautious. It restrains the individual from such activities that may harm him. However, fear may also act as a spur to greater activity. Fear of loss of prestige may motivate an adolescent to accomplish a task at any cost. Fear acts as a check on behaviour. Fear of consequences makes an individual hold his behaviour within reasonable limits. Fear guides an individual to fashion his life in such a way that it would be appreciated by the community in which he lives. Also fear helps an individual in the attainment of a variety of social virtues."There are values that are served by anger. It is used as attack in attempts at overcoming fear. An individual may be jarred out of his lethargy by its use", Write Crow & Crow (1964, p. 94). A person by being aroused to anger may improve his ways and overcome states of indecision. He can proceed to deal constructively with a problem.

Children can express their thoughts, wishes or criticisms under the spur of anger. This would provide valuable insights for their elders. An angry child is a person struggling with a problem. Anger may help him to dispose of the problem. Soon after a moment of anger, if a person asks himself: "Why did I get so mad?" then, anger can be turned to useful purpose. Such an inquiry, if sincerely pursued, may often teach a person much about himself. "Probably the most important factors in a child's emotional development are the affection that he receives *from* and his opportunity to develop wholesome affection *for* his fellow creatures.

An attitude of affection—the love that a parent feels for

a child—and similarly the love of a teacher or any other adult for his pupils or the children with whom he deals—is likely in one way or another to leave its mark on everything that goes into the day-to-day business of living", states Jersild. A teacher is a parent substitute. Every child under the care of a teacher seeks a degree of "warmth" or fondness in his relations with his teacher. An accepted child feels free to try things out, to explore, to make mistakes and in the process, to learn. A rejected child, either at home or school or both may take recourse to a variety of defence mechanisms. He may turn aggressive, may try to withdraw from the threatening situation or become regressive in his behaviour. A child feels "secure" if he loves and is loved. Therefore, affection is important for an individual's emotional welfare. Each child needs to feel that he is wanted and liked, that he belongs and has value to others. Confidence in the fact that others are concerned for him is an important factor in promoting a child's "security".

In the school if a child feels rejected, he may turn sullen, restless and a regular absentee. He will show his unwillingness to learn and even to the extent of destroying school property. A high level of morale in the school will help the children in growing up well emotionally. The teachers should be intelligent, sympathetic and democratic in spirit.

7

THEORETICAL APPROACHES OF COGNITIVE DEVELOPMENT

The degree of mental alertness of an individual influences every phase of his life pattern. Growth of mental ability helps the individual to adapt his behaviour to environmental conditions as the individual strives to achieve increasingly conscious goals or purposes. Throughout his life, an individual has to react to stimulations that he receives from a multiplicity of elements in his physical environment. He has to sensitively respond to many and varied stimulations that he receives from his human interrelationships. The more sensitive and adequate his responses are to these stimulating conditions, the more successful he is in his daily life activities. Growing intellectual ability aids emotional maturity of children. Greater and better understanding of situations, objects and persons results in their changed emotional reactions to these.

Intellectual growth reduces the number of situations arousing the emotions of anger and fear in them. Fear, anger, and even love change as understanding develops. Intellectual growth is an imperative for social development, learning and adjustment. His mental alertness enables a child to detect if he is accepted or rejected by the group. If learning is a problem-solving activity, it is an impossibility without mental ability. For wholesome adjustment, an intelligent understanding of the surrounding objects, events, persons and situations is a must.

There are two major ways of describing children's cognitive development—"a stage approach" and "a process approach". In this chapter we shall discuss the former and in the chapter that follows, the latter. The Swiss psychologist Jean Piaget has made the most extensive study of children's cognitive development.

After many years of careful observation, Piaget has developed a theory of how cognition develops through a series of stages as children mature. "Any book published within the last twenty-five years dealing with children's cognitive development will have a considerable section on the work of the great Swiss psychologist Jean Piaget. His work has been enormously influential and has sparked off numerous studies throughout the world. He is above all concerned with the qualitative changes that take place in children's thinking as they move from infancy to late adolescence", remark Siann and Ugwuegbu (1985, p. 41).

Siann and Ugwuegbu have made three very important points about Piaget's work in general:

(i) "According to Piaget a child does not think *like* an adult only rather less effectively. Instead, he believes children's thinking to be *qualitatively* different from the thinking of the adults.

(ii) As a biologist Piaget traces cognitive development in terms of biologically inherited ways of interacting with the environment.....

(iii) Piaget tries to identify the thinking structure of each age level and he tries to show they become adapted by a dynamic interaction with the environment. He then tries to show how these structures combine together to form more and more powerful systems of thought. Thus we see that his approach is open in which each stage builds up in a progressive way on the proceeding stage. Further, the thinking characteristic of later stages is more complex than that of the earlier stages because of the environmental experience that has been gained and because of the maturation that has taken place" (*Ibid*, p. 43).

Basic Theoretical Concepts

A brief explanation of some *basic theorectical concepts* in Piaget's system is necessary before any attempt is made for a faithful description of the major periods of intellectual development.

(1) *Schemes*. Piaget uses the term "*schema*" in the singular. The plural of the term is either "*schemas*" or "*schemata*". The term "*schema*" refers to anything that links, or organises, or ties together. It means something like "*arrangement*". The activities that people perform on objects are known as schemes. Schemes are what can be repeated and generalised in particular acts—the sucking scheme and the pushing scheme may be treated as examples. For Piaget, the development of intelligence is the result of the increasing ability of the child to co-ordinate his actions, both physical and mental. These abilities or capabilities are the child's schemes. In the early development of the child the schemes are physical actions on the objects in his immediate environment. As the child moves towards adolescence, these actions become internal mental operations.

"The concept of organisation and adaptation, called "functional invariants" by Piaget, are essential to this theory of the development of human intelligence. They are constants in the functioning of human intelligence and cause the development of cognition to proceed in the manner that it does. These mechanisms operate continuously. Through organisation we arrange and systematise perceptual and cognitive data into units and patterns that have meaning for us. When for example, we see a rectangle at the far end of a room, we perceive a door, and know by which route we can leave. Through adaptation we can alter the ways in which we deal with the needs of the environment by changing ourselves or by changing the environment" (Gibsan, 1976, pp. 51-52).

(2) *Assimilation*. Adaptation according to Piaget, involves two processes: assimilation and accommodation. Dececco & Crawford (1977, p. 71) have stated thus : "The child needs the environment in order to develop his intelligence. Intellectual assimilation is similar to biological assimilation. Using the example of food, Piaget (1970, p. 707) wrote: 'From a biological point of view, assimilation is the integration of external elements into evolving or completed structures of an organism. In its usual connotation, the assimilation of food consists of a chemical transformation that incorporates it into the organism". "In its more general sense, the word '*assimilation*' means to make similar, or to ignore a difference, or to eliminate a difference, or to treat things as if there were no differences between them", writes Stephens & Evans (1977, p. 66). We *assimilate* when we take in a new stimulus from the environment and respond to it with a behaviour we have already used for a familiar, and in some way similar stimulus.

The evolving or the completed structures are the scheme. Assimilation is the process of extracting from the environment what is needed for developing and maintaining schemes. When a child encounters an orange for the first time and rolls it like a ball, he is assimilating.

(3) *Accommodation*. Piaget (1970, p. 708) defines *accommodation* as any change of a scheme by the elements it assimilates. The accommodation of the new food substances changes the existing digestive schemes. The respiratory structure in the new born infant enables him to take his first

breath. The total experience of breathing modifies these respiratory structures. If the infant does his breathing at sea level, he will develop one kind of lung structure. If he lives high up in the Himalayas, he would acquire somewhat different lungs. Because of these changes or accommodations in the respiratory structures, he can now carry on additional kinds of activity. These activities in turn shall produce further changes or accommodations. In other words, the progressive modification of schemes through accommodation allows the child to develop his capabilities beyond the point of dealing with the immediate physical environment. We *accommodate* when we add a new activity to what we already know or when we modify an old behaviour. A child knows that a ball rolls; when he learns to bounce it, he has added a new behaviour.

(4) *Equilibration.* For Piaget the development of intelligence is not the passive unfolding of experience and heredity. But the continuous interaction between maturation and experience—between the structures the child possesses and the world around him—is the key to development. At any time, the child will have inherited or developed some structures for dealing with the world that he has experienced so far. So long as these prove adequate, he has established some sort of harmony; or equilibrium with the world. When the child comes across a different experience, this experience has to be maintained or a new equilibrium has to be acquired. This process of maintaining or regaining equilibrium is called *equilibration.*

Equilibration is the process of seeking mental balance. It is the process that produces progressive equilibrium between assimilation and accommodation. In other words, to deal with any kind of disequilibrium, either real or threatened, the child has two mechanisms at his disposal—the mechanism of assimilation and the process of accommodation. Equilibration is a dynamic function. It moves development from simple to more complex schemes through the dual action of assimilation and accommodation. Piaget suggests that development is a progressive equilibration from one plateau, as it were, to another.

The Developmental Periods

According to Piaget's theory, the child's intellectual life may be blocked off into four main stages—each stage having a number of substances and marked by the attainment of a definite equilibrium. A stage usually defines as set of behaviours that occur together. As a group. They characterise a quality of behaviour that differs conceivably from the quality of behaviour in earlier and later stages. These stages are separated by transition periods. Stages follow each other in an orderly sequence and each stage builds up in a progressive way on the proceeding stage. The transition periods involve a process of integration, whereby the behaviour from the earlier stage is transformed into the next, along with some new elements. Further the thinking characteristics of later stages is more complex than that of the earlier stages. This results on account of two factors. The first is because of the environmental experience that has been gained; and the second is because of the maturation that has taken place.

1. *Sensorimotor Stage.* This stage lasts from birth through to approximately two years of age. The ages given are averages. They may vary considerably, depending upon intelligence, cultural background and socioeconomic status. During this stage language plays little part, but there takes place sensorimotor linkages—schemata or schemes—between the world of sensations on the one hand and the world of action on the other. In other words, during this stage takes place the internalisation of the sensorimotor schemate.

In the first half of this period, the child's activity is centred on his own body. In the second half, the child develops schemes of practical intelligence that enable him to deal with objects in space. During this period through countless experiments, the infant learns to separate himself from the external reality.

At birth, the infant displays a limited number of unco-ordinated reflexes. During the next four months these are co-ordinated into simple schemes. For example, during the first month of his life, the infant is passive to the world of objects. The newborn infant grasps passively but he does not reach actively for any object. For the first four months in his life, the world of objects is still only an extension of the infant's needs and movements. However, new response patterns are formed by chance from combinations of primitive reflexes. The baby's first accidentally finds its way into his mouth, through a co-ordination of arm moving and sucking.

During the next four months (4-8 months), new response patterns are co-ordinated and repeated intentionally in order to maintain interesting changes in the environment. Infants move their feet to shake the dolls, hanging from the hoods of their bassinets. An infant learns that if he shakes a rattle, it will make a noise. At about eight months of age one important discovery is the concept of object permanence. The infant is able to realise that objects outside himself continue to exist even though he cannot see them. For example, if a cloth is placed over a toy for which a seven-month-old is reaching, he immediately stops his motion, makes no efforts to search for it and acts as if the toy ceased to exist. In contrast, a nine-month old infant will actively search for an object that has been hidden under a cloth or behind a screen. He seems to realise that the object exists even though he can no longer see it. "In this stage the child begins to develop primitive notions of space, causality, and time and shows the beginnings of imitation"

At the fourth sensorimotor stage (8—12 months, the behaviour of the infant becomes even more intentional and active. During this period there is development of more purposive behaviour. The baby actively searches for vanished objects. If a watch is hidden under a cushion, the baby searches for the watch and finds it. But even at this age his search is limited. If the child is allowed to watch the watch being transferred to another location and hidden under a second cushion he would continue to search for the watch under the first cushion only even though he has watched while it was being concealed in a new location.

At the fifth sensorimotor stage (12-18 months), there is more active experimentation by the infant. He is able to carry out goal directed behaviour. He experiences causality, as what happens when he pushes his food dish to the edge of the table. The baby achieves object constancy. He consistently looks for an object where it was last seen to disappear regardless of what has happened on previous trials. By the end of this stage (1½ -2 years) the child is able to produce new solutions to problems through mental explorations. He can imagine actions as well as execute them. There is internalisation of sensorimotor behaviour patterns. These sensorimotor schemata serve as the first symbols to the child. The infant gets great pleasure from the symbolic manipulation of the absent object. They enrich his experience and refine it. During the sensorimotor period the infant is egocentric to a considerable degree.

(2) *Preoperational Stage.* Between the ages of two and seven, approximately, the child is in the preoperational period. The infant begins to think symbolically. He links up the crude sensorimotor schemata, the existing ones, with the more precise names or concepts made possible by language. Thus, his internal cognitive picture of the world grows. Though the child is able to represent reality as language and mental imagery, his thinking is full of contradictions. For the child it is the period of magic. He treats a doll as though it is a live baby, or a block of wood as though it is a car. During this period, words, pictures, emotions, fantasies and dreams, all seem part of an external

reality. He sees everything from a single point of view, his own. His thinking is unstable and *egocentric*. The child can see the world only from his own standpoint. He is completely dependent on his own viewpoint. Therefore, the child has difficulty in taking the viewpoint of others. During this period assimilation appears to outweight accommodation.

During this period the child learns to classify objects by single salient features. For him if A is like B in one respect, it must be like B in other respects. He soon learns for example, to put rectangular blocks in one pile and triangular blocks in another. These are graphic collections. According to Piaget, graphic collections are not true classifications. Piaget and his co-workers have found that the child may place a triangle over a square, because he thinks that these two forms must be related. The triangle reminds him of the roof of a house while the square can be the main part of the building. Therefore, according to his logic at this stage, the triangle must be placed over the roof of the building and nowhere else. The child can use the word "dog" for one set of objects and the word "bird" for another.

Mastery of classes does not come easy. The child may visualise a class in some fashion but still has trouble in seeing its limits, and seeing its relation to things outside. When he is asked: "If someone were making a collection of parrots, could he add other birds to his collection?" The child's answer would be in the affirmative. When asked, "Why?" His answer would be: "Because parrots are birds". The preoperational child has difficulty in dealing with two systems of classification at the same time. At one time he can decide whether a ball is big or small. At another time he can decide whether it is red or blue. But he finds it hard to sort big red balls into one category and small blue balls into another.

Toward the end of this period the child begins to develop *conservation concepts* and begins to use numbers. As adults we take conservation principles for granted. For example, the mass of a substance does not change when its shape is changed or when it is divided into parts, or the liquids do not change in quantity when they are poured from a container of one shape to that of another. For children, however, attainment of these concepts is an aspect of intellectual growth requiring several years.

A five or six year child has not yet attained the conservation of numbers. It is difficult for him to see that number is an attribute that can be separated from the particular objects to which it is applied. The child watches ten coins put together to form a cluster. Then he watches while someone scatters the coins over a wider area. He now thinks that there are more coins than before. Later on, in the first phase of the concrete operational period, this misconception will disappear. While he is about eight years old, the child will clearly realise that the number of objects is unaffected by the area they occupy.

A child younger than seven has difficult with conservation concepts because his thinking is still dominated by visual impressions. Examine the following experiment with continuous quantity. A child is shown two identical glasses filled with the same amount of coloured water. When asked if the glasses have equal amounts of water, the child says that they do. The experimenter then pours the content of the first glass into a teller, thinner glass and asks the child to compare the amount of water in the third glass (the taller and the thinner) with the amount in glass number one. Any child who is less than 6½ years old will say that glass number three has more liquid than glass number one.

The preoperational stage is divided into two subperiods:

(1) *Preconceptual* period 2-4 years) and

(2) *Perceptual* or *Intuitive* period (4-7) years). In the former assimilation outweighs accommodation and

in the later accommodation outweighs assimilation. The preconceptual period is dominated by symbolic play. At this sage, the child builds sand castles and uses sticks as guns. The intuitive period is dominated by imitation. The child reproduces the movement of monkeys in the zoo and he models the behaviour of parents and older brothers and sisters.

(3) *Concrete Operational Stage.* The concrete operational stage falls between the ages of seven and twelve. Piaget calls this period the concrete operational stage because, although the child is using abstract terms, he is doing so only in relation to concrete objects.

The main characteristics of this period are that the children are able to engage in activities of the mind which Piaget calls "*operations*". Operations are mature mental routines and their acquisition is the heart of mental growth. Also operations are actions on objects that can be carried out mentally without being carried out physically. These operations are also *reversible.* They can be followed by another mental act which restores objects to their original stages. Changing the appearance of an object does not change the amount of it. This is an *operation.*

The amount of clay remains the same if some clay made into a ball at first is rolled out into a long sausage shape. Next. In the preoperational, stage, the child makes the chemical scales balance only by adding some weight to the lighter pan - A. For him, if he added too much; he then added something to the other pan. During the stage of concrete operations, he comes to see clearly that as far as the *effect* on balance is concerned, adding something to pan. A is precisely the same as taking something away from pan B. During the concrete operational period the child begins to think in terms of a set of interrelated principles rather than in single bits of knowledge. He thinks in terms of systems.

The child thinks more logically. Now he believes that number, length, mass, weight and volume remain constant even if superficial changes are made in their external appearance. As he works his way through this stage, the child attains *conservation concepts*—first number, then mass and about the age of nine, he comes to conserve weight. And towards the end of the elementary school stage, he is able to conserve volume.

During this period, at first the child realises that the number of objects in a group remains constant or invariant no matter whether someone seatters them over a wide area or gathers them together in a compact bunch. The clay in the longer object is equal in amount to that in the round object, out of which the former was made or, out of two balls made of an equal amount of clay, keeping one as the reference ball and rolling the other into a longer object. He can see that weight as an attribute remains the same for two substances though differing markedly in appearance, as in the case of the clay ball and the longer sausage-shaped clay. Finally, he sees that the volume of a given quantity of liquid remains invariant no matter whether it is poured into a shallow, broad glass or a tall, slender glass.

During this period, the child understands the relationships among various concrete operational groups. *Grouping"* refers to logical operations. Groupings are intellectual schemes which possess particular characteristics, namely, (a) closure, (b) associativity, (c) reversibility, and (d) identity. According to Berlyne.

(a) *Closure*: Any two operations can be combined to form a third operation.

Examples : 5 + 3 = 8; A is five miles south of B and B is seven miles south of C. Therefore, A is 12 miles south of C; all men and all women = all human adults.

(b) *Reversibility*: For any operation there is an opposite operation which cancels it.

Examples : 3 + 4 = 7 but 7 - 4 = 3; A is five miles south of B and B is 7 miles south of C. Therefore. A is 12 miles south of C; but A is 12 miles south of C and C is 7 miles north of B. Therefore, A is five miles south of B.

(c) *Associativity*: When three operations are to be confined, it does not matter which two are combined first. This is equivalent to the possibility of arriving at the same point by different routes.

Examples. (3+4) + 5=3 + (4+5); all vertebrates and all invertebrates = all human beings and all subhuman animals; A is the uncle of B and B is the father of C=A is the brother of D and D is the grandfather of C.

(d) *Identity*. There is a 'null operation' formed when any operation is combined with its opposite.

Examples: 2 – 2 =0; all men except those who are men=nobody; I travel 50 miles to the north and I travel 50 miles to the south = I find myself back where I started.

Children are able to construct groupings in the concrete operational stage. These groupings are mental structures. They help children to join classes to form larger *classes*, and to multiply classes. One kind of grouping involves combining classes to form larger classes. The larger classes can be divided into subclasses through a reverse operation.

According to Piaget, at this stage children are more able to reason and think about *classes* and sub-classes which they failed to do while at the preoperational period. Thus, if he is shown eight green beads and four white beads, all made of glass, he is able to answer correctly that the child in the previous stage cannot: "Are there more green beads than glass beads?" The child in the preoperational period tends to confuse "glass" and "white" and answers that there are more green beads.

The same thing can be explained in a different way "Children in the concrete operational period also understand that an object may fit into two sets at once. Piaget once showed a group of five, six and seven year olds a box that held twenty white and seven brown wooden beads. He asked each child in turn whether there were more white or brown beads. All got the answer right. Next Piaget enquired: "Are there more white or more wooden beads?" The younger children, still preoperational, replied: "There are more white than brown beads". They could not comprehend that beads could be both wooden and white or brown. The concrete operational children responded; "There are more wooden than white beads because all of the beads are wooden and only some are white". That kind of reasoning is called *decentration,* because children have become able to consider two or more attributes of an object at the same time; they no longer centre on a single attribute".

In the concrete operational period the child is able to think logically but he cannot think in abstract logical terms. This is the difference between this stage and the final period of cognitive development. At the concrete operational period the child can order objects on the basis of a dimension such as height or weight.

He can also form a mental rapresentations of a series of actions. An eight year old can readily draw a map of the route to a friend's house which a five year old cannot. But at this stage though the child uses abstract terms, he does so only in relation to concrete objects. At this stage he cannot generate and test a set of logical hypotheses. In the final stage of cognitive development the youngster is able to reason in purely symbolic terms.

Formal Operational Stage. This stage is from 11-12 years through to adulthood. In this stage the child begins to think logically. He can think in abstract terms. He can follow logical propositions, reason by hypothesis. He begins to appreciate that some hypothetical problems can be solved in your head, by applying the same rules as would be applied to concrete problems. He can isolate the elements of a problem and systematically explore all possible solutions. Whereas the younger child tends to deal largely with the present, the child in this final stage becomes concerned with the hypothetical, the future and the far away.

The major characteristic of this period concerns the possible versus the real. The adolescent theorises about and criticises the present operations of the world because he conceives of many possible ways in which it could operate and many alternative ways in which it could be better.

There are two typical tests to illustrate the difference in the thinking of children in the periods of concrete and formal operations. The first typical experiment is that a child is shown five colourless, odourless liquids in test tubes. Out of these, the first tube contains diluted sulphuric acid; the second tube contains water; the third one contains oxygenated water; the fourth one contains this sulphate; and the fifth one, here '*g*', contains potassium iodide. A dropper is also presented with bottle '*g*' to the child.. Then he is asked to discover what combination of the five will produce a yellow liquid. The child in the concrete operational period attempts to do it by trial and error. But the child in the stage of formal operations goes about the problem systematically, for instance, combining the first and second, then first and third, and so on, till 1+3+ 'g' yield a yellow colour.

In the second typical experiment the child is presented with a length of string suspended from a hook and several pendulum bobs of different weights that can be attached to the lower end. He can vary the length of the string, change the attached weight, and alter the height from which the bob is released. Now the child is asked to try to discover what determines the period of oscillation of a simple pendulum.

It is observed that a child still in the concrete operational stage experiments changing some of the variables, but not in a systematic way. An adolescent on the other hand, sets up a series of hypotheses and proceeds to test systematically each one. He is able to reason that if a particular variable, say weight, affects the period of oscillation, then the effect will appear only if he varies one variables only, and holds all others constant. If this variables does not seem to have any effect on the time of swing, he rules it out and tries another.

As Deceaco & Crawford, (1977, pp. 93-94) have stated, three other characteristics of the period of formal operations are: First, the adolescent's thinking is basically hypothetic—deductive. Second, thinking in this period is propositional thinking. Third, thinking in this period involves combinatorial analysis. An Hilgard and his coauthors (1976, p. 77) remark that considering all the possibilities, working out the consequences for each hypothesis, and confirming or denying these consequences is the essence of what Piaget calls formal operational thought.

Before their evaluation of Piaget's theory, Siann and his coauthor (1985, p. 45 comment that throughout the child's progress through these stages, the child is seen as organising and adapting to his environment in a way that depends upon his biological maturation, his past experience and the particulars of his own situation.

A number of different psychologists in recent years, have directed mainly three sorts of criticism on Piaget's work.

(1) Children show the ability to be logical at ages well below those Piaget specifics.

(2) Children can be trained to solve conservation experiments correctly without actually physically handling the types of material used in the experiment.

(3) Experimental findings show that children often failed Piagetian tasks not because they were incapable of the reasoning required but because the experimental conditions confused them.

However, for the educator the following important points emerge from Piagetian theory.

(1) Cognitive development occurs in interaction with the environment. Therefore, the educator should make the classroom a stimulating environment with plenty of material for the child to see, handle and hear because, the more stimulating to the classroom environment, the more dynamic is the children's cognitive development.

(2) Symbolic thought does not always occur using language; and hence, sometimes cognitive development may not be dependent on language. Some children think in images and relationships that are not directly dependent-upon language. Therefore, teaching conducted largely through the verbal medium may not always be appropriate for all children. In this circumstance, the wisest course of action an educator should take is to provide the children a number of very stimulating games designed to encourage thinking and reasoning in ways that are not heavily dependent on the development of language. Through these games the teacher can encourage children to think analytically and constructively.

On the positive side, Piaget's research constitutes the most monumental body of fact and theory centred on the cognitive development of children.

The Implications. It is possible for as to deduce certain implications from the following fundamental factors of the cognitive theory of Piaget.

"From infancy to adulthood, the child is assimilating data from the world around him, accommodating to the world around him, and organising his knowledge and experiences into a system or structure. Piaget stresses this active organising as the most important force in the child's cognitive development. Other recognised forces are: maturation, experiences with the environment and direct teaching by other people, *Equilibration* is the term Piaget gives to this active internal processes of organising and co-ordinating one's own intellectual development. Through the process of equilibration -a form of self-regulation—the organised system, or cognitive structure, is constantly changing, enlarging, and reorganising as more experience with the world and with himself presents the child with ideas and events that must be assimilated or accommodated to" (Klausmeier & Ripple, 1971, p. 407).

For Piaget, cognitive development is not an automatic process. Assimilation, accommodation and equilibration are merely the equipment that makes cognitive development possible. Assimilation and accommodation occur only when an environmental stimulus is appropriately matched with the particular level of cognitive development attained by the child. A child's interaction with his environment becomes meaningful to his development only when assimilation and accommodation occur equally and continually. A balance must exist between the two processes. They must be in *equilibrium*. Development is a progressive equilibration. Environmental stimulation increases the development of the child's ability to solve problems. Each stage of cognitive development has a particular form of equilibrium as a function of the problem-solving abilities of that stage. Equilibration allows the transformation from simpler to more complex conceptual thinking to the final equilibrium represented by the mature adult mind. One cannot lose sight of the fact that cognitive development results out of biologically inherited ways of interacting with the environment and that biological forces are exceedingly important in the emergence of periods of cognitive development.

Thus, it is the job of the teacher and the school to provide the opportunity for rich and varied experiences to the children under their care. But the experiences must be suitable to the child's period of cognitive development "Teaching children concepts that they have not attained in their spontaneous development...is completely useless," Piaget has cautioned (Hall, 1970, p. 30). You can tell kindergartners that among the industrially developed countries, Japan ranks second. They can parrot the information back to you. The fact means nothing to them. On the other hand, toys of different geometrical shapes, even mechanical toys, presented to the children to play with and games involving activities that help the co-ordination of arm-hand-finger and vision, leading to writing, drawing, colouring, threading and carving would be appropriate and meaningful. The subject-matter to be presented to the child must correspond to his level of cognitive development. For example, a child in the sensorimotor stage of development should be provided with experiences that are action-linked, whereas a child in the formal-operational stage may be provided with more theoretical experiences that match his ability to cope with complex and abstract problems.

Preschool education gives the child an opportunity to act directly on concrete objects and has a particular connection with the development of language. As the child is developing his ability to use language, the teacher must be aware of the child's level of thinking and speak with him at this level.

Thus, matching the subject matter with the child's level of cognitive development relates to providing concrete experiences and the concomitant development of language. The teacher, therefore, should select stimuli that his students can use and understand; otherwise, the performance of his students shall be poor. They shall feel frustrated or bored.

In selecting classroom activities the teacher is playing an exceedingly important role in influencing the cognitive development of the child. A teacher who provides stimulation that is appropriate to the child's cognitive level is indeed helping the child to learn to solve problems at his level of thinking, and therefore, the child can proceed to the next level. Piaget stressed interaction with the environment for the processes of assimilation and accommodation can occur only in a stimulating environment. If at any time a teacher ceases to provide new stimuli in the child's environment, both processes would stop and the development of the thinking processes would thus be curtailed.

For teachers one of the problems is that cognitive development cannot accurately be measured in terms of chronological age. Children of the same chronological age may be behaving on different cognitive levels. As Piaget is very particular about it, cognitive development gets heavily influenced by biologically inherited ways of interacting with the environment; as a result of which children of the same age group might be having different levels of cognitive development. Because biological forces are extremely important in the emergence of periods, children and growing youngsters enter each period of cognitive development at varying ages. Children from different backgrounds enter the same class. Some might have come from very rich and stimulating environments and others from impoverished environments that do not provide sufficient or appropriate types of stimulation.

For example, Ramesh, a ten-year old, enters a particular

class. He is in the concrete operational period of development. But teachers observe him having difficulty with conservation concepts and often with *decentration*, that is, ability to consider two or more attributes of an object at the same time. These difficulties hint that preoperational period of development still lingers in him. In this case, the only explanation a teacher can offer is that Ramesh probably has the capability of concrete operational thought; but because of, perhaps, an impoverished environment, Ramesh cannot *demonstrate* some of the characteristics that show concrete operations.

Since children of the same chronological age may be behaving on different cognitive levels; some educators therefore, feel that children should be given the freedom to work individually at tasks of their own choosing. But the learning situation should not be unstructured. The teachers must provide the materials of interest that suit each one individually. This will also suit children from different backgrounds who do not solve problems in the same ways.

Presentation of new events interest children. This is an aspect of environmental stimulation that educators have learned. "When a new event is both familiar enough so that it may be assimilated without distortion into the students current cognitive level, and novel enough so that it produces some degree of conflict, then interest and learning are promoted. The child is "ready" to learn". "The disparities in behaviour in a single individual reflect the experience or lack of experience the individual has had in an area. The area may be either educational, in the limited sense of curricular, or social. A youngster may get all. As in school yet fail socially out of insufficient contact with other children and with adults. Learning from others -by what psychologists call "social transmission" is a major factor in development".

In Piagetian theory, cognitive growth proceeds with spurts and plateaus of achievement; and hence, a curriculum must be designed with this in mind. A child is not a miniature adult. He is qualitatively different with a distinctive mental structure. Teaching strategies must take this into account. The child should be helped to develop his full intellectual potential to become an adult. Hence it doesnot mean that children growing up in familypatterns and role descriptions differnet from those described above are in any way at a disadvantae. definitely is that all children grow up in families i.e., among their parents and siblings and frequently also their grandparents, each of whom plays an important role in their development.

8

CHILD PSYCHOLOGY AND SCIENTIFIC METHODS

Educational psychology uses the methods of science as its primary "tools" of investigation. It has certain things in common with all the sciences to the extent that educational psychology uses scientific methodology. Art is a skill or knack for doing something which is acquired by study, practice and special experience. Science can be defined both as an *organised body of knowledge* and as a *method and system of deriving truth*. The latter is certainly the more crucial aspect since it not only permits the discovery of knowledge but also affords us the means for accelerated scientific progress. Therefore, Kendler states: "Science is a method designed to discover and interpret knowledge based on rules of evidence and logic."

However, the *basic aim of science is theory*. The basic aim of science is to explain phenomena, natural or man-made. Such *explanations* are called *theories*. Instead of trying to explain each and every separate behaviour of children, the scientific psychologist seeks general explanations that encompass and link together many different behaviours. The teacher may not go on explaining children's methods of solving problems in mathematics. He may seek general explanations of all kinds of problem-solving. He might call such a general explanation a theory of problem—solving.

It is the job of scientific enquiry to discover those conditions or factors that bring about, determine or cause a particular event. Science can be said to have four purposes: (1) to *describe,* (2) to *predict,* (3) to *control,* and (4) to *explain.*

All sciences have as a goal the *accurate description* of some aspect of the natural universe—description of some natural or even man-made phenomenon. The psychologist, of course, has chosen the behaviour of man and other animals as his sector of the universe. Behaviour is what organisms do: how they act and what they experience. Descriptive investigations also involve the discovery of new phenomena. Many important scientific discoveries have been made in the process of describing events or structures. The descriptive function of science depends upon instrumentation for measurement and observation; for example, the use of electron microscope by the biologist and the standardised tests of intelligence by the psychologist are quite essential for description of phenomena of interest to these scientists. Advancement of science is not possible without identification and accurate description of objects, events, persons and situations.

Description provides the basic knowledge hat is necessary for realising the other purposes of science. One of these purposes is *prediction,* which is the ability to predict phenomenon X from phenomenon Y. For example, the next stage of an embryo's development can be predicted very accurately from knowledge of the embryo's current stage. A student's achievement in school can be predicted fairly accurately by an aptitude test administered a year or two earlier. To make a scientific prediction we must first identify a consistent relationship between persons, events, situations, and conditions. The ability to predict is an important step toward explaining phenomena. Observations of association between phenomenon X and phenomenon Y may arouse the curiosity of the scientist about why the association occurred. This will lead to the development of a theory or an urge to collect further data about the phenomena.

The third purpose of science is to gain *control* over the phenomena being studied. Scientific experiments are designed to achieve this purpose. In an experiment, phenomenon X is manipulated to determine whether it exerts control over phenomenon Y. The control purpose of science is closely related to its predictive purpose. The major difference is in whether phenomenon X is manipulated by the scientist. In predictive enquiry the scientist studies naturally occurring associations between phenomena. In experiments, the scientist manipulates phenomena to determine whether he can control the condition of some other phenomenon. Experimental manipulation leads to more powerful statements about associations between phenomena than does predictive enquiry. Experimental results do permit conclusions about cause and effect relationships which predictive associations between phenomena necessarily do not. In the field of psychology during the past few decades, the ability to plan and direct many aspects of human behaviour has increased considerably. In the area of vocational guidance, for example, psychologists have been able to determine how success and satisfaction in a particular occupation are related to specific patterns of abilities, interests and other measurable human characteristics. Utilising this knowledge, the vocational counsellor is able, through personal interview data and psychological test scores, to offer advice as to the vocational fields that are most likely to offer success and happiness for a particular client. Such control is used widely by the armed forces of U.S.A. and other European countries in classifying men for special training.

The fourth purpose of science -*explanation*—is considered to be its ultimate goal. If scientists are able to explain a set of phenomena, it means that they can describe, predict and control the phenomena with a high level of certainty and accuracy. Scientists have always been eager to understand the phenomena they investigate. Scientific understanding of an even occurs when the event is explained. The function of

Science is to discover those factors which determine the occurance of an event. This is essentially what an explanation is -the specification of those factors or conditions related to the occurance of an event. Higher order explanations consist of general statements that apply to many different events having some feature or features in common. Everybody observed that apples fell, that stones tossed high up ultimately came down, and that the waters of the seas periodically rose and fell, and also that planets followed certain paths in their trip around the sun. Galileo and Newton saw a basic similarity among these events. The theory of gravitation, that is, the attraction that masses of matter have for each other, was formulated to apply to all of these formerly "*isolated*" events. Thus, a theory is a statement or a group of statements the purpose of which is to explain a group of events. The theory of gravitation, like all scientific theories serves two major functions. In the first place, it *integrates* existing data, and in the second place, it predicts new events. To conclude scientific *explanation* takes the form of a *theory* about the phenomena being investigated.

Theories and Language of Research

Scientific explanation takes the form of a theory about the phenomena being investigated. Scientific theory is defined as "a system for explaining a set of phenomena by specifying constructs and the laws that related these constructs to each other". And also as: "*A theory is a set of interrelated constructs (concepts), definitions and propositions that present a systematic view of phenomena by specifying relations among variables, with the purpose of explaining and predicting the phenomena.* The definition of scientific theory by Kerlinger says three things. "One, a theory is set of propositions consisting of defined and interrelated constructs.

Two, a theory sets out the interrelations among a set of variables (constructs) and in so doing, presents a systematic view of the phenomena described by the variables. Finally a theory explains phenomena. It does so by specifying what variables are related to what variables and how they are related, thus enabling the researcher to predict from certain variables to certain other variables". An educational psychologist might have a theory of school achievement in mathematics. His variables might be intelligence, socio-economic status, personality adjustment, motivation and spatial relationships and numerical aptitudes. The phenomenon to be explained is school achievement in mathematics. School achievement in mathematics is explained by specified relations between each of the six variables and school achievement in mathematics, or by combinations of the six variables and school achievement in mathematics. The educational psychologist successfully using these set of constructs, then "*understands*" school achievement in mathematics. He is able to "*explain*" and to some extent at least, "*predict*" it.

In this section, rather in the discussion carried on so far in this chapter, certain terms that are quite new for you have been used. Any scientific discipline has need for a specific language for describing and sumarising the observations in that area. Scientists need terms at the empirical level in order to describe particular observations. Also they need terms at the theoretical level for referring to hypothetical processes. The scientists use terms that are labels for concepts and constructs, both at the descriptive and theoretical levels. The following are some of the general terms used in educational and psychological research.

Concepts and Constructs: A *concept* is a class of stimuli which have common characteristics. These stimuli are objects, events or persons. We ordinarily designate a concept by its name, such as book, tree, liquid or war student or dedicated teachers. A concept is a word that represents the similarities or common aspects of objects or events that are otherwise quite different from one another. Words such as '*tree*' and '*liquid*' represent common aspects of otherwise diverse things. The purpose of a concept is to simplify thinking by including a number of events under one general heading. Some concepts are quite close to the events they represent. Thus, for example, the meaning of the concept *'tree'* can easily be illustrated by pointing to specific trees. The concept is an abstraction of the characteristics that all trees have in common—characteristics that are directly observable.

Constructs are created to summarise observations and to provide explanations. Terms such as *'motivation'* or *'problem-solving ability'* cannot be easily illustrated, by pointing to specific objects, events or persons. These are higher level abstractions referred to as constructs. From the concepts, scientists have constructed these complex abstractions. Constructs are created by combining concepts and less complex constructs in purposeful patterns. As a scientific construct, "motivation" means both more and less than it may mean as a concept. As a construct, it can be used in two ways. One, it enters into theoretical schemes and is related in various ways to other constructs. For example, we may say, *"school achievement* is in part a *function of intelligence* and *motivation."* Two, *"motivation"* is so defined and specified that it can be observed and measured. We can observe the motivation of children by administering a properly designed test on them or we may ask the teachers to tell us the relative degrees of motivation of their pupils. We can combine concepts such as *'visual acuity,' 'symbol discrimination,' 'listening ability,'* and others, in a purposeful pattern to produce the construct *"reading readiness."* Constructs are useful in interpreting experimental data and in building theory. They are used to account for observed regularities and relationships.

Concepts and constructs should be defined in abstract terms, which give the general meaning they are supposed to have and in terms of the operations by which they can be measured in the particular study. The former type of definition is called the *constitutive* definition and the latter is known as the *operational* definition. A *constitutive definition* refers to a more formal type of definition in which a construct is defined by using other constructs. For example, "*intelligence*" may be defined as "*the ability to think abstractly.*" In this definition we have substituted one concept for another. This type of definition helps to convey the general nature of the phenomenon in which the investigator is interested as well as to show its relation to other theories and studies using similar concepts.

A constitutive definition elucidates a concept or construct and perhaps, gives some insight into the phenomenon described by the term. "All constructs, in order to be scientifically useful, must possess constitutive meaning", says Torgerson. This means that they must be capable of being used in theories. An *Operational definition* is one which assigns meaning to a concept or construct by specifying the operations that must be performed in order to measure that concept, construct or variable. This type of definition is essential in research since data must be collected in terms of observable events.

Constructs like '*anxiety* or '*achievement*' must represent some kind of observable events for any empirical study. To define a term operationally, scientists must choose discriminable events as indicators of the concepts and devise operations to obtain data relevant to them. Intelligence can operationally be defined as the scores on X intelligence test, or intelligence is what X intelligence test measures. An operational definition of creativity may refer to scores made on the Minnesota Test of Creativity. An operational definition is a sort of manual of instructions to the investigator. Operational definitions permit the researcher to find a relationship between two sets of observable and measurable data that were selected to represent the concepts, constructs or variables. For example, an investigation of the relation of constructs—creativity and intelligence—in practice will relate scores on an X intelligence test to scores on an X measure of creativity.

It reflects or expresses some concept for construct. A variable takes in different values. It varies in the same individual from one time to another; and it varies between the individuals at the same time and between averages for groups. Intelligence, Achievement, Sex and Social class are all variables. While height and weight are physical variables, intelligence and anxiety are psychological variables. Educational psychologists are concerned in determining how such variables are related to one another.

The concepts and constructs that are of interest in a study become variables. Some variables are *antecedent* to other variables. For example, to be able to read, one must have some degree of intelligence. Here, the variable intelligence is antecedent to the variable reading ability and to some extent, the latter is a *consequence* of the former. The level of reading ability is dependent on the level of intelligence. In research, Variables that are consequences of or dependent upon antecedent variables are called dependent variables and the variables that are antecedent to dependent variables are called *independent variables*. Independent variables predict the *dependent variables*. In experimental research, a change in the dependent variable brings about a corresponding change in the dependent variable under study.

Scientific Methods

Jut as the teacher does in working through an immediate practical problem in the classroom, the scientist, even on the highest level, observes, generalises, makes inferences, observes further and then accepts, revises or rejects his generalisation. In other words, scientific method is a "problem-solving activity" and especially, "a hypothesis generating and testing sequence." The teacher may use the essence of the scientific method in dealing with a problem in the classroom. A student, say, 'R' entered the eighth class a month ago. The teacher notices that 'R' obtained the lowest score in the last arithmetic test, in the class. (This is the initial observation of the teacher.)

He then checks back and discovers that 'R' has completed his homework in arithmetic only twice during the past month, has obtained falling scores on both tests given, and has not answered to oral questions during this time. (For the teacher, these are the observed regularities.) Then the teacher conjectures : "Perhaps, 'R' is mentally retarded." (This is the hypothesis.) If so, 'R' must be doing poorly in other subjects. (This is the further inference or deduction.) The teacher checks the records and finds that 'R' is doing better than average work in geography and science and is only slightly below average in everything else, except translation work. (These further observations discredit the hypothesis which is then discarded.) Can 'R' be emotionally disturbed and fearful in the new school? (Thus, the teacher formulates a new hypothesis.) If so, it should show up in all school subjects (new inference). The records have already shown that this is not so (hypothesis discarded). Perhaps "R" has a specific disability in arithmetic (third hypothesis). If so, a diagnostic test should be helpful (deduction). A standardized diagnostic arithmetic test showed that 'R' was unable to do any sum on 'fractions', but could do almost all other arithmetic processes fairly well (hypothesis supported). If the trouble of "R' is of this specific type, teaching him the 'fraction' processes should bring about improvement (further deductions from the hypothesis). Providing remedial work and following up on the consequences will lead to either acceptance, rejection, or modification of this hypothesis. And, this is a "hypothesis generating and testing sequence."

"The scientific method is commonly thought to include three major phases," say Borg and Gall. "The first step is to formulate a hypothesis, which is a tentative proposition about the relation between two or more theoretical constructs....The next phase of the scientific method is to deduce empirical consequences of the hypothesis....The Third phase of the scientific method is to test the hypothesis by collecting data....According to the scientific method, the hypothesis will be either supported or rejected by this empirical data."

After intellectualising the problem, after turning back an experience for possible solutions, after observing relevant phenomena, the scientist may formulate a hypothesis.

A hypothesis is a conjectural statement, a tentative proposition, about the relation between two or more phenomena or variables. Our scientist will say, "If such-and-such occurs, then so-and-so results."

Observation

It should be clear by now that the observation-test-experiment phase is only part of the scientific enterprise. If the problem has been well stated, the hypothesis or hypotheses adequately formulated and the implications of the hypotheses carefully deduced, this step is almost automatic—assuming that the investigator is technically competent. On the basis of the research evidence, the hypothesis is accepted or rejected. The information is then fed back to the original problem, and the problem is kept or altered as dictated by the evidence.

In any science, several different methods of study must sometimes be applied to a problem before all the pieces fit together properly. Each method contributes in its own way to the final solution. In planning a study, the scientist attempts to select the method or methods most appropriate to the particular problem under consideration. The scientist has to collect empirical data to test the hypothesis he has formulated. On the basis of the research evidence, the scientist accepts or rejects the hypothesis. Kerlinger tries to include all the methods of investigation under his caption. "Observation-Test-Experiment." We shall now consider some of the methods by means of which the educational psychologists obtain their data.

The five chief methods that the psychologists use are the experimental methods, the observational method, the survey method, the text method and the case study method. These methods are not mutually exclusive and a variety of observational and measurement techniques are used in each.

Experimental Method

The essense of the experimental method is simple. The experimenter

(i) changes or varies something,

(ii) keeps other conditions constant, and

(iii) looks for an effect of the change or variation upon the system he has under observation. Since educational psychology is the science of human behaviour in educational situations, the experimenter looks for an effect of the changes he has made upon behaviour of the subject or subjects. This is simple enough, but becomes more and more complicated as we consider the experimental method is greater detail.

The experimental method can be used outside the laboratory as well as inside. The distinguishing characteristic of a laboratory is that it is a place where the experimenter can carefully control conditions and take measurements in order to discover orderly relationships among variables. The ability to exercise precise control over variables distinguishes the experimental method from other methods of observation. In every experiment we must have at least one independent variable and one dependent variable. In simple experiments, such as the one described below as an example, there is only one variable of each kind.

But it is possible with appropriate experimental design to have several independent variables and several dependent variables in the same experiment. A teacher (the experimenter) seeks to discover whether learning ability depends on the amount of sleep a pupil has had. The teacher can control the amount of sleep by arranging to have several groups of subjects spend the night in the school laboratory. One group might be allowed to go to sleep at 11-00 P.M., another at 1.00

A.M., and the third group might be kept awake until 4-00 A.M. By waking all the subjects at the same time and giving each the same learning task, the teacher-experimenter can determine whether the subjects with more sleep master the task more quickly than those with less sleep. In this study the different amounts of sleep are the antecedent conditions, the learning performances are the results of these conditions. Thus, in this study the amount of sleep is the *independent variable.* It is independent of what the subject does and the teacher can manipulate this variable learning ability in this experiment is the *dependent variable.* It is affected by charges in the antecedent conditions or the independent variable. The dependent variable is a measure of the subject's behaviour which the experimenter does not manipulate. The experiment described here can be clearly and simply expressed by the formula: $y = f(x)$. In this formula y represents the dependent variable or the learning performances of the subjects; f represents the phrase "is a function of "in the sense of "is causally dependent upon"; while x represents the independent variable, the antecedent condition, that is, the different amounts of sleep, upon which the occurance of y depends. Thus, for the experiment under consideration we could say that the subject's ability to master a learning task is a function of the amount of sleep he has had.

A very important characteristic of the experimental method is *control.* In an experiment it is important that only the variables be allowed to change. Factors other than the independent variable which might affect the dependent variable must be held constant. It would do no good to study the effects of varying an independent variable if other factors, unknown to the experimenter, changed also. In an experiment we must control conditions which will give misleading results. Because of the greater difficulties in controlling experimental conditions in the behavioural sciences of which educational psychology is one, and also education is included, experimental research is very difficult to apply to certain, kinds of educational problems. The typical experimental design, often called the *before-and-after experimental design*, used in educational research involves: (1) the selection of a sample of subjects; (2) random assignment of these subjects to experimental and control groups; (3) the *exposure* of the *experimental group* to a *treatment* that is withheld from the *control group*; (4) and, finally, the evaluation of the two groups on the dependent variable, that is the variable the teacher-experimenter is attempting to change. A simple experimental study described below shall clarify these steps.

Suppose, we want to prove the efficacy of Problem-Development method over all other methods of teaching Geography. The first step we should take is to select a sample of 200 pupils from a particular class and randomly assign 100 pupils to the experimental group and 100 pupils to the control group, after administering on the subjects an achievement test in Geography as a pretest. The next step for us is to teach Geography using the Problem-Development method to the experimental group and to the control group by any traditional method. This process continues for about one year. At the end of one year, as our next step in experimental research, we administer an achievement test in Geography on both the experimental and control groups, as the *post-test*. Now we have the required data on hand for statistical computation to determine whether the subjects who had been exposed to the treatment, that is, taught Geography by the Problem-Development Method, have obtained more favourable scores on the achievement test given as the post-test, than those students who comprised the control group. If the subjects of the experimental group scored higher than that of the control group, it would prove the efficacy of the Problem-Development Method over other methods, to some extent. This is also called the *pretest-posttest* experimental design.

Observational Method

The most basic technique in psychology, as in any other

science, is observation, But *incidental observation* of naturalistic behaviour is neither scientific nor psychological. Of course, the early stages of a science necessitate exploration to become familiar with the relationship that will later become the object of more precise study: but, incidental observation of naturalistic behaviour is seldom a useful source of data or inference. Incidental observations are typified by the anecdotal accounts of the behaviour of one's pet dog or cat or of one's own children. Reports of the behaviour of wild animals by naive and untrained observers are also included in this class.

Because psychology is the science of behaviour and mental processes, some scholars are of the opinion that there are two kinds of observations which have been employed in psychological research – *introspection* and *objective observation*. It has already been mentioned in Chapter 2 of this text that Wundt, Titchener and their followers used the technique of *introspection* to analyse conscious experience. The psychologist himself, being a trained introspectionist, observed his own mental processes and furnished information in the form of verbal reports. The behaviourists doubted the validity of this method. It could not be employed to analyse the conscious activities of animals and children. However it gets employed in the form of verbal reports in an objective manner uptill how. Hilgard et al., (1979, p. 18) have referred to Masters and Johnson who in their extensive study of the physiological aspects of human sexuality, developed appropriate methods for observing and recording the responses of the subjects in the laboratory. The data included (1) observations of behaviour, (2) recording of bodily changes, and (3) responses to questions asked about the subject's sensations before, during and after sexual stimulation, which means that the subjects made a verbal report of their conscious experiences in so far as their sexual response was concerned.

The second major technique psychologists use to obtain facts about behaviour is *objective observation*. Here the experimenter notes what his subject does—not what he feels or thinks. The environmental conditions under which a response occurs, as well as the response itself are systematically described. Systematic and objective observations of people, usually children, under natural conditions can be very informative. Objective observation is mainly concerned with what goes on in the external world. However, the psychologists of this age have many instruments for observing what goes on inside the organism. Among these are devices for studying the electrical currents of the brain and the changes in conductivity of the skin during emotions. Introspection is aimed at describing inner experience and objective observation at describing aspects of behaviour observable to others. Thus, some psychologists think, they supplement each other, and often they are used in combination. An example of this combined technique is an experiment conducted to investigate the ability of blind people to orient themselves in space.

In educational research, observational method is given a very important place. It is treated as basically a measurement technique of gathering data about the subjects involved in a study. Of course, the techniques employed in observational methods are different to a very great extent form other measurement techniques. Observational methods permit the collection of direct information about human behaviour that can be collected only indirectly measurement techniques such as paper and pencil tests.

Observational methods help the investigator to study in detail specific areas of human behaviour.

For example:

(a) how do preschool children respond to a television programme that contains a high number of violent acts?

(b) What specific counselling procedures are most highly related to realistic vocational decisions by high school students?

The tools generally used in observational research are:

(a) Standard Observational Schedules,

(b) Audiotape and Videotape Recorders,

(c) Rating Scales,

(d) Sociometric techniques

(e) Anecdotal Records and the like. Observational methods to be successful, observation of behaviour must be made by trained observer or observers; observational information should be properly recorded; the observed behaviour should be objectively defined and the data collected should be analysed as completely and as quantitatively as possible.

Survey Method

Some problems are very often difficult to study by direct observation. Also very often it is not feasible or desirable to spend a long period of time in the field or to obtain data about all the individuals in a group. In such circumstances, much information, useful and valuable, may be obtained by conducting a survey. In this method the investigator, by using written questionnaires or oral interviews, obtains data from a selected group. It is important that he chooses his group carefully so that it will constitute a *representative sample* of the larger group or *universe*: in which he is interested. A representative sample is one drawn in such a way that every person in the universe from which it is drawn has an equal chance of being selected. If the sample is not representative, the data collected will be of little value. Therefore this method is known as "*Sample Survey Method.*" The best known type of survey are the "opinion poll," and "census."

Survey research in education typically employs questionnaires and interviews to determine the opinions, attitudes, preferences, and perceptions of persons of interest to the investigator. Survey studies that deal with sensitive topics such as "premarital sex relations" frequently use interviews. Surveys in education can be used to explore a very wide range of topics such as the extent to which open classrooms are being used at the elementary school level and the perception of teachers and principals about the effectiveness of such classrooms; and procedures currently used to help handicapped children adjust to regular classrooms problems encountered by first year elementary school teachers also can be taken for survey study.

Text Method

The tests and scales are important research instruments in contemporary education and psychology. They are used to measure all kinds of variables—achievement, intelligence, aptitude, interest, attitude, and other kinds of abilities and accomplishments. Tests and scales are measuring instruments. However, *while tests are scales, scales are not necessarily tests*. Scales do not ordinarily have the meanings of competition and success or failure that tests do. Significantly, we say "achievement tests" and "*attitude scale*." Most of the hundreds, perhaps thousands, of tests and scales can be divided into the following classes: intelligence and aptitude tests, achievement tests, personality measures. Tests enable the investigator to obtain large quantities of data from people without elaborate laboratory equipment. A test essentially presents a uniform situation to a group of people who vary in aspects relevant to the situation. An analysis of the results then relates variations in test scores to variations among people.

A coefficient of correlation is an alternative to experimentation. Suppose, a teacher wishes to find out the extent to which "intelligence" affects the "achievement" of pupils. His hypothesis is "intelligence influences Achievement." He wants to test this hypothesis. Treating "intelligence" as the independent or predictor or X variable and "achievement" as the dependent or criterion or V variable, he administers an intelligence and an achievement test on a sample of pupils drawn from the population. The, he scores both the tests and computes a coefficient of correlation between the two sets of scores. He then proceeds to estimate the regression of X on Y and the regression of Y on X. With the help of these regression coefficients the teacher is able to tell us the extent to which "intelligence" affects or predicts "achievement," that is, statistically put, if values of Y variable are predictable from values of X variable.

Tests and Scales are indispensable for educational research, particularly in causal-comparative method or export facto research, correlational studies and evaluation research. The study of relationships between educational variables is extremely important. The causal comparative method is one approach for exploring relationships between variables. The correlation method provides another approach for exploring relationships. Like causal—comparative approach, the distinguishing characteristic of correlational studies is the way in which the relational data are analysed. "Educational evaluation is the process of making judgement about the merit, value, or worth of educational programmes, projects, materials and techniques". The causal comparative method is often used instead of the experimental method because many of the cause and effect relationships that we wish to study in education do not permit experimental manipulation. It should be emphasised again that Tests and scales are very essential for all these types of research.

However, test construction and use are no simple matters. They involve many steps in item construction, item analysis, scaling, establishing norms and so on and so forth. Later chapters may explore the problems of testing in a very brief manner. Some authors call the case-study approach as life-

History Method. In the case study approach, the investigator studies the conditions in people's lives which have led to their becoming whatever they are, be they physicians, teachers, politicians, alcohol edicts or even criminals. The investigator wishes to find out the causes of a specific form of behaviour by studying the antecedents. But according to some authors, the case study in its simplest form, involves an investigator who makes a detailed examination of a single subject or group or phenomenon.

Including only the clinical method under case-study approach, Sawrey and Telford proceed to say: "The clinical method is not primarily a research method, it is rather a set of procedures for the diagnosis and treatment of people with behaviour problems," Including case-study approach under observational research, Borg and Gall write: "Several kinds of case studies can be found in behavioural science literature. These include:

(1) Historical case studies of Organisations

(2) Observational case studies

(3) Oral histories

(4) Situational analysis

(5) Clinical case study.

"The three basic life-history methods are the daybook method, the clinical method and the biographical method'. States Ruch. This text follows Ruch. The *daybook method* is often used in child study. The child's development is carefully observed and recorded day by day. Accumulated records of this kind provide us standards by which we are able to judge the normality of the development of individual children.

The starting point of a *clinical investigation* is an individual who needs and/or seeks help and the procedure ideally terminates with the better adjustment of the individual. Thus, the clinical method is ordinarily used when and only when people come to the psychologist with personal problems. Not all clinical problems require thorough study. But when they do, the psychologist usually begins by getting a detailed account of the past life history and present life situation of a person in an effort to discover the cause of and solution to some emotional or social adjustment problem. This information is usually gained by interviewing the person and his associates. The clinical study to be most effective needs three professional functions. A physician examines the individual for signs of physical illness; a specially trained social worker examines the patient's home conditions, social background and environment; and a psychometrician administers and scores various tests of aptitude and personality. The clinical investigator may seek the help of these professionals to obtain the required data. When a full record has been, obtained of the past and present influences in the patients' life, the information is put together, discussed and steps are taken to diagnose the problem, treat or remedy the difficulty. The tests, the diagnosis and the remedy, of course, will differ from case to case.

The *biographical method* is simply an attempt to obtain psychological understanding from an analysis of the records of men's likes as set down by themselves or others. Scientific biographies are important sources of data for psychologists studying individuals. "Most case histories are prepared by reconstructing the biography of a person according to remembered events and records. Reconstruction is necessary because the person's earlier history often does not become a matter of interest until he develops some sort of problem at such time understanding of the past is thought to be important to comprehension of present behaviour." There are certain important limitations to the scientific value of this method.

9

TRENDS IN PHYSICAL GROWTH AND IMPORTANCE STUDY

As teachers we should know what the human child is made of and how he grows and matures. Unless we know the structure, we can't know the function. If education can be defined best as a process of bringing about changes in people, then the study of physical growth is a necessity; the reason being " "Growth in intellect, personality and character can be adequately understood only in relation to physical growth". A child's mental and personality development, and his success in school, depend to a large extent upon his health, energy, rate of growth and general physical fitness. Knowledge about a child's physical nature may be revealing in studying his behaviour in his home and at school. "Only in relation to the physiological changes going on in childhood and adolescence can we interpret changes in interest and attitudes during this period". The physical appearance of a child is of much importance to him. The child whose level of physical development is at variance with that of his friends' may feel insecure and ashamed. Any physical abnormality exerts as tremendous influence upon a young person's emotional reactions.

A sound knowledge of physical development and the

normal variations may help the teacher in guiding the social and personal adjustment of the child and the adolescent. It may also provide a base for curriculum construction, determination of teaching procedures and proper utilisation of learning equipment.

Physical Growth after Birth

A number of studies in child development have resulted in definite conclusions concerning trends of physical growth after birth. The general growth trend of the body is pictured as follows.

(a) Growth is rapid in Infancy (from birth up till age two).

(b) It slows down in early and middle childhood (from ages two to eight);

(c) An increased growth rate begins between the ages 8 and 10 and continues till adolescence, that is, ages 17 and 18.

(d) From this time on, there is a tapering off of growth to maturity (ages 24 and 25). As educators, we are mostly concerned with child development from birth to maturity. Adulthood or maturity—from ages 25 to 60—can be regarded as the period of life during which the individual can utilise whatever powers and potentialities he has achieved during his earlier stages of development. Old age starts after 60 during which the powers and strength gradually decline.

There are two aspects of physical development—Anatomic and Physiologic Anatomic refers to skeletal growth, including the growth of bones structure, changes in height and weight and changes in body proportions. Physiologic includes developmental changes in the nervous, circulatory, respirative, digestive and excretory systems and also muscular and glandular systems. Life depends upon the proper functioning of these systems.

Changes of Height and Weight

Eighty-five percent of the new-born infants weigh between 6 and 9 pounds and vary between 17 and 22 inches in height. On the average, a boy is about 20.5 inches in height and 7.5 pounds in weight whereas a girl is about 20.3 inches and 7 pounds. The weight of an infant doubles in the first 6 months. The height of an infant increases by about 50 percent during the first year of life and the weight triples. The rate of growth gradually decreases from the second year till age 12 or 13.

Upto puberty, the male child is larger in all body measurements than the female child. The girls reach puberty when they are about 11 to 12 years old, about two years earlier than boys; and during this period the girls surpass the boys in total size. However, the growth spurt of the adolescent girls is less intense and shorter than that of the adolescent boys. When they are about 14 years old, boys begin to grow faster that girls in height and weight. The average age at which the adolescents reach adult height is approximately 16 for girls and 18 for boys. A youngman is about 4 inches taller and about 25 pounds heavier than a young women. From ages 6 to 18, during these 12 years, boys increase over one-half in height and over 3 times in weight. Adolescent girls prefer the company of adolescent boys, older, stronger and bigger than them as they feel very secure and happy in the company of these boys. Ultimately this results in younger girls carrying older boys.

There are wide variations in height and weight for men and women. In height men may range from the little dwarf to the tall giant and in weight from the midget to the fat man of the circus. Norms for height and weight in Indian conditions are lacking. Sufficient research has not yet been conducted in this area. However, boys and girls of India are less tall and weigh less than their counterparts in European countries. In our country, standard height and weight charts for use in schools and colleges are not available. However, teachers should encourage children to have physical exercise and good food for an excellent body build. Undernourishment and improper functioning of endocrine glands result in stunted growth which may cause emotional disturbances in children, particularly adolescents. When teachers come across such cases, they may advise the parents to consult a doctor for treatment to bring about some improvement in the physical build of these children. Teachers should be aware of the fact that relative height and weight are necessary in so far as they are indicative of good physical health and do not deviate greatly from the average making the child emotionally upset. Indians do not bother much about height and weight and I am an Indian.

Changes in Body Proportions

Physical growth is from the head downward and from the trunk outward. Infants, children, adolescents and adults differ in their body proportions. A new-born infant's head is approximately one-fourth of its total body length and that of an adult is about one-tenth. The trunk is about three-eights the total length of the body at birth which remains approximately the same at adulthood. At birth, the legs are three-eighths the body length but at adulthood they are one-half. From infancy up to puberty, head growth is slow, limb growth rapid and trunk growth intermediate. As the child grows older, while the proportion of the head decreases, that of the legs increases correspondingly. As the years pass by, all the parts of the body grow larger; but at adulthood, the arms and legs are proportionately longer.

Body contours also change with increasing physical maturation. The infant has a small flat nose which grows and becomes prominent when he reaches adolescence. Before the onset of puberty, difference in body contours between boys and girls is very little. When they reach the early years of adolescence, their body contours become significantly different. Adolescent boys have broad shoulders, narrow hips and straight leg-lines. On the other hand, girls have narrow shoulders, wide hips and curved leg-lines. Gradually, the bodies of boys and girls get resculptured into that of young men and young women.

Anew-born infant has 270 bones. They are small, spongy, pliable and loosely connected. He has soft spots in his skull which usually get closed before the end of his second year When he is about 13 to 14 years old, the human individual has 350 bones which have become hardened, with increased mineral content and changed chemical content During the process of osification (hardening of bones), some of the smaller bones fuse together. Therefore, at adulthood, the number of bones is reduced to 206. The bony structure of human beings follows a relatively regular growth pattern but on account of ossification and nutrition, individual differences in the bony structure are occasionally noticed. In connection with skeletal growth, reference also may be made to dentition. The infant is born toothless but gradually the first set of "baby teeth" appear. As the child grows older, the "baby teeth" fall one by one and get replaced during childhood by a second set of "permanent teeth". With proper care and good food, the second set of teeth serve the individual throughout life.

Physiological Development

The growth of the skeletal structure and the internal organs of the body are closely related. Both develop from birth onward; but the internal organs mature at different rates. In its rate of growth, each internal organ may vary from any or all others.

The musculature system increases both in weight and in its power to control the skeletal movement. Muscles constitute 23 to 25 percent of the total body weight in infancy, 27 per cent at age 8, 44 per cent at age 16, and in an adult about 45 per cent. In general, men have greater muscular power than women.

The heart is 5 times as large at age six as at birth. From age six to puberty, it grows more slowly. From age 12 to age 18, within six years, it doubles in weight and volume. During the embryonic and fetal periods, there is evidence of frequent but feeble heart beat. Respiration starts after birth. During the first month of life, the heart beats 140 times per minute. As its size increases, the heart beats more and more slowly. The blood pressure in the arteries increases. At age six, the rate of heart beat is 100 times per minute, and at age 12, eighty-five times. For an adult, the normal rate of heart beat is about 72 times per minute.

At birth, the infant possesses his full number of neurons (nerve cells). The neural structure is approximately what it shall be in future life. The neurons strengthen rapidly. The nervous system reaches its final size by the time the child is 12 years old. The maturation of certain nerve cells in the brain makes it possible for certain mental developments in later life. The lymphatic system eliminates body wastes and destroys bacteria. It increases rapidly for the first 12 years of life. The activity of the lymphatic system decreases rapidly after this age.

The function of the endocrine glands was discussed earlier. Different endocrine glands grow at different rates. The thymus and the pineal glands are very active during childhood and do not allow sex glands to mature early. These two glands turn less active during adolescence. The thyroid and pituitary glands grow faster compared to other glands.

In the life of a boy or girl, sexual maturity is of great importance. The genitals grow very slowly during childhood. They being to grow rapidly with the onset of puberty and the reproductive cells—sperm and ova—get secreted in the testes and ovaries of boys and girls respectively. Also the six glands secrete hormones that develop the secondary sex characteristics in boys and girls. Girls begin to menstruate when they are about 12 to 13 years old. During the years 12 to 14 their breasts develop, hair grows, on the public triangle and the facial contour changes. The boys mature later than girls, the average age being 14. In the boys, hair grows in the public region and the arm pits, the shoulders broaden, the voice deepens, beard grows on the face, the facial contour changes and the sweat glands under the arm pits become active and odorous. As menstruation in girls, in the boys nocturnal emissions follow the development of secondary sex characteristics. Those bodily movements that result from the coordinated functioning of nerves and muscles are usually referred to as moter activities.

Motor development is necessary for the survival of the human child. Motor skills serve individuals in their day-to-day living. Motor activities involve physical strength, the movement of gross and finer muscles, in well co-ordinated fashion and dynamic strength that causes the whole body movement. Gross movements of the body are needed in such activities as running and bicycling.

Dynamic strength is needed for such activities as swimming and gymnastics. Writing, painting, the use of different kinds of tools and appliances need finer muscular coordination. Movements after birth are mainly reflex actions and largely automatic. With the nervous system and musculature becoming mature, the child gains control over different motor functions. Maturation determines the rate and pattern of early motor responses. Later motor development depends on both maturation and practice, experience and training. Motor growth takes an orderly sequence. Motor control begins in the head region of the infant.

At first it tries to hold its head erect and fixate the eyes. Next, it develops precision in reaching for and grasping objects that interest it. For the first month of life the baby lies on its back, at the end of which it begins to lie on its stomach; It is difficult for the infant at this stage to change its position from the prone to the sitting of from the sitting to the prone posture. By the fourth month it sits with support and by the seventh month, sits alone and erect. When it is 8 months old, it stands up when helped, at the 9th month stands up holding an article of furniture, at the 11th month walks when led, pulls itself up to stand by furniture when it is 12 months old, at the 13th month tries to climb up steps, stands alone by the 14th month and walks alone when it is 15 months old. The infant begins to creep when it is about 8 to 10 months old.

Individual differences are noticed in the method of locomotion. Some infants crawl on all fours while others crawl with their buttocks on the floor. Some infants prefer crawling to walking even after they have learned to take steps. With some encouragement they shift from crawling to walking. Those infants who crawl early, are likely to learn walking also early. Given good health, play space and sufficient number of toys, the infant shall exercise its muscles to gain motor control. Normal motor development may get hampered on account of physical defects and ill health. Parents should not try to hasten the individual maturational motor process. They should allow the infant to develop at its own rate the motor coordinations.

Physical strength is estimated from the strength of the grip. Ability to grasp depends both upon physical strength and muscular coordination. The grasping movement of the infant is a reflex and is feeble. When the baby is about one year old, grasping with the thumb and fore finger begins and continues through adulthood.

At first, the movements of the infant are awkward and fumbling. It is unable to manipulate the objects in the physical environment well. His arm and hand movements develop in a consistent pattern through the first six years of life of the child. Gradually the infant begins to manipulate the play things, drinking equipment and other objects. This helps the child to develop specific skills in gaining control of objects by the time he enters into formal schooling.

In the elementary and secondary schools, much time and attention is given in developing motor skills in children. They are engaged in a variety of activities that demand correct motor responses. Some of the most important activities are: penmanship, drawing dancing, cooking, sewing, laboratory experiments, and various games. All these activities need motor control. Strength, speed, precision, rhythm, timing and finer muscular co-ordination are the most important aspects of motor execution. Long periods of training and practice are needed to develop these abilities in children. During the practice periods, errors are eliminated and correct responses are established.

Not only intense practice but a desire to succeed and a confidence that successful performance can be attained are also essential for the development of motor skills. There are wide differences in strength of grip among children. Variations in strength of grip are directly related to differences in ages at which puberty occurs and the accompanying increases in weight appear. Boys are found to be superior to girls in strength of grip.

This superiority increases after age 14 and affects the whole personality of adolescent boys and girls in a variety of ways. There are a number of motor activities such as dancing, playing different games and stenography, which depend upon the co-ordination of and integration of smaller muscles. Co-ordination of movements is necessary to learn handwriting. When they are about six years old, children are able to control movements better. From this time on, improvement in motor control and regularity of movement makes legible handwriting possible.

Speed of response is needed for hitting or kicking a ball. Rhythm is closely associated with music and dancing. Timing is important for such complex responses as striking a golf ball or hitting a base ball. Playing Cricket is an excellent example. It involves all these aspects of motor execution. Any single act we perform involves muscular co-ordination. If the muscles and nerves do not work together in proper combination, we do not get the result aimed at. From age six onward improvements in the finer motor co-ordinations take place in children. They learn games involving running, throwing, hitting, kicking and balancing in space. Boys are superior to girls in playing games which involve running, hitting and kicking. In motor performances that depend upon finger dexterity and flexibility of wrist movement, girls tend to be superior to boys. Therefore, girls are found to dress themselves earlier than boys and show proficiency in such activities as knitting and needle work and stenography.

Dynamic strength stands for that strength which moves the whole body in a defined direction. Most of the outdoor games such as swimming and gymnastics depend much upon the dynamic strength of the individual. Activities such as climbing and broad jump also need dynamic strength. Boys are found to be superior to girls in the display of dynamic strength and perform such activities as broad jump, distance throw and vertical jump much better than the girls. Biological differences, differences in cultural expectation, motivation and practice are perhaps responsible for the widening differences between adolescent boys and girls in such activities as running, jumping and throwing.

Some individuals are right-handed, some others left-handed and still some others who are right-handed for certain activities are left-handed for others. It is not possible to tell the proportions of right-handed, left-handed and neither right nor left-handed individuals. There are many theories to account for the handedness. Some think the left-handed are born as such. Others claim that it stems out of the chance use of left hand in the early years of life. Still some others explain it as a result of the dominance of the right hemisphere of the cerebulum. Also there are others who think that it is on account of the training given in using the left hand in the early years of life. Therefore, there is no universally accepted position about handedness. Teachers may encourage children to be right handed but on no account try to change handedness if it is likely to upset the child emotionally or may result in the child feeling insecure. Even though they are left-handed, there are individuals who are as dexterous and efficient as their right handed cuterparts.

Educational Significance of Physical Growth

The most important, helpful, deceiving, confusing and neglected fact in education is growth. Most important, because the first requirement through the school years should be that healthy growth be achieved; most helpful because the plasticity and vigour of the growth period greatly aid education; most deceiving, because education frequently mistakes as its own accomplishment what is actually the result of maturing; most confusing, because growth involves constant change; most neglected, because growth is so usual and yet so complex and often subtle." Physical growth influences learning, personality and adjustment of children.

Some of the physical and sensory defects and disorders that adversely affect the learning and adjustment of both children and adolescents and present a lot of psychological problems are defective vision, defective hearing and speech organs, physical deformities, chronic illness and malnutrition. A dominant motive in the life of an adolescent is to be like others and gain approbation of both youngsters and adults. Any shift in physical proportions which seems to the adolescent boy or girl to make him or her different from others, plunges them into an ocean of worries. It is the social implication of any physical anomaly that is of much importance. It is the attitude of others toward such physical conditions that affects the emotional balance of the adolescents.

Physical defects and deviations, therefore, present a lot of psychological problems for the teachers to resolve. Besides physical deviations and sensory defects, there are other hazards the children meet with, Children meet with accidents. They are severely burned or run over by automobiles or drowned. If they survive, they are confined to the hospital for a pretty long time and come out of the hospital, invalid, disabled, disfigured or with mental aberrations, Interruption of school work, unwholesome personal-social relationships, and a lot of other troubles creep in on account of these incidents. Illness and accidents are so common during childhood and adolescence that they have to be reckoned with. Physically handicapped children deserve special attention and extra help. Their learning difficulties have to be diagnosed and determined. Remedial teaching to meet the nature and extent of their learning difficulties should be given to them. The negative attitude they have developed should be slowly but surely removed and self confidence should be built up in them. Materials which are interesting and in the right proportion to the children's readiness should be collected and used. Close attention should be paid to these children and necessary guidance in new teaching–learning situations should be given to them.

They should be given early opportunities to use and demonstrate their newly learned skills. However, children, who are receiving remedial teaching, should not be engaged in any competition. It may undermine their growing confidence. Physical disorders such as unusual fatness, thinness and malnutrition can be rectified by diet control, proper nutrition, exercise and rest.

Poor food habits, strenuous exercise and fatigue should be avoided. Good nutrition has good effects on the temperament of the child. As safety education for children is necessary to help them avoid fatal accidents, health education is necessary to form in them good habits of health and hygiene and also a positive attitude towards health practices. Children should be encouraged to abide by the rules of the road and form other desirable habits in their day-to-day living. Constant medical check-up of the boys and girls would not only reveal their health condition but also their personal problems. For the treatment of chronic illness, parents should be advised to consult doctors.

There are certain physical indispositions and sickness that are psychological in origin. They result from the fruistration of psychosocial needs in children. Teachers should try to find out' the causes of such ailments and try to reduce the emotional stress the children suffer from by being sympathetic towards them. The teachers should try to make children realistic and self confident. Most of speech disorders such as stammering and stuttering arise from psychological stress. Teachers dealing with such children should provide them with situations in which their sense of worthwhileness would be built up.

Dynamic strength integrated with motor control and co-ordination have a very high value in our civilisation. Athletic ability has a prestige value in the present-day society. Children who are unable to participate in different games on account of their inadequacies feel themselves insecure and inferior as a result of which their emotional-social adjustment gets affected.

Therefore, a physical education programme in the schools is a necessity. The primary aim of any physical education programme in a school should be to develop physical fitness in children and should be conceived as a series of planned developmental experiences.

Athletics, too strenuous is not conducive to optimum physical growth. Competitive sports are too severe for those who are at the formative period of life. Therefore, everything should be "moderate" for children and adolescents. A well-trained coach should be there in all schools. He should lead the children from participating in milder sports to strenuous games. Each child should be well acquainted with the game he could play throughout his life Physical handicaps should be engaged in activities adapted to their capacity and needs. An understanding of physical growth and change throughout the school years is basic for an adequate understanding of the development of abilities and personality in children through these years. There are four important cycles of physical growth after the infant is born.

10

PSYCHOANALYTIC LITERATURE IN THE ADOLESCENCE

Thinking at adolescence and afterwords reaches its most natural form. The adolescent is now capable of solving problems is abstraction. The subject of adolescence after an interval of twenty years. During this time much has happened in analytic work to throw added light on the problems concerned and to influence the conditions of life for young people, whether normal or abnormal. Nevertheless, in spite of partial advances, the position with regard to the analytic study of adolescence is not a happy one, and especially unsatisfactory when compared with that of early childhood. With the latter period, we feel sure of our ground, and in possession of a wealth of material and information which enables us to assume authority and apply analytic findings to the practical problems of upbringing.

When it comes to adolescence, we feel hesitant and, accordingly, cannot satisfy the parents or educational workers who apply for help to us and to our knowledge. One can hear it said frequently that adolescence is a neglected period, a stepchild where analytic thinking is concerned. These complaints, which come from two sides, from the parents as well as from the analytic workers themselves, seem to me to warrant closer study and investigation than they have received so far.

Psychoanalytic Literature In the Adolescence

The psychoanalytic study of adolescence began as it well known, in 1905 with the relevant chapter of the Three Essays on Sexuality. Here, puberty was described as the time when the changes set in which give infantile sexual life its final shape. Subordination of the erotogenic zones to the primacy of the genital zone; the setting up of new sexual aims, different for males and females; and the finding of new sexual objects outside the family were listed as the main event. While this exposition explained many features of the adolescent process and behaviour which had been unexplained before, the newly developed notion of the existence of an infantile sex life could not but lower the significance of adolescence in the eyes of the investigators. Before the publication of the three essays, adolescence had derived major significance from its role as the beginning of sex lie in the individual; after the discovery of an infantile sex life, the status of adolescence was reduced to that of a period of final transformations, transition and bridge between the diffuse infantile and the genitally centered adult sexuality.

Seventeen years later, in 1922, Ernest Jones, London, published a paper on "some Problems of Adolescence" which dwelt on a "correlation between adolescence and infancy" as its most distinctive point. Following the pronouncement in the Three Essays that the phase of development corresponding to the period between the ages of two and five must be regarded as an important precursor of he subsequent final organisation, he showed in detail how "the individual recapitulates and expands in the second decennium of life the development the passed through during the first five years..." He ascribed the difference in "the circumstances in which the development takes place" but went as far as propounding "the general law...that adolescence recapitulates infancy, and that the precise was in which a given person will pass through the necessary stages of development in adolescence is to a very great extent determined by the form of his infantile development". In short: "these stages are passed through on different planes at the two periods of infancy and adolescence, but in very similar ways in the same individual".

Jones's important but isolated contribution to the problem coincided with a peak in the publications of Siegfried Bernfeld in Vienna, a true explorer of youth, who combined work as a clinical analyst and teacher of analysis with the unceasing study of adolescence in all its aspects of individual and group behaviour, reaction to social influences, sublimations, etc. His most significant addition to the analytic theory was the description of a specific kind of male adolescent development (1923), the so-called "protracted" type which extends far beyond the time limit normal for adolescent characteristics, and is conspicuous by "tendencies toward productivity whether artistic, literary or scientific, and by a strong bend toward idealistic aims and spiritual values..." As the solid background for his assumptions, Bernfeld published, in cooperation with W. Hoffer, a wealth of material consisting of self-observations of adolescents, their diaries, poetic productions, etc.

While Bernfeld accounted in this manner for the elaborations of the normal adolescent processes by the impact of internal frustrations and external, environmental pressures, August Aichhorn, also in Vienna, approached the problem from the angle of dissocial and criminal development. His work lay with those young people who answer to the same pressures with failure to adapt, faulty superego development and revolt against the community. His book *Wayward Youth* (1925) acquired world renown as the outstanding pioneering attempt to carry psychoanalytic knowledge into the difficult realm of the problems of the young offender.

Based on familiarity with S. Bernfeld's views, and intimately connected with A. Aichhorn's studies, I contributed in 1936 two papers under the titles "The Ego and the Id at Puberty" and "Instinctual Anxiety During Puberty." In my case, interest in the adolescent problems was derived from my concern with the struggles of the ego to master the tensions and pressures arising from the drive derivatives, struggles which lead in the normal case to character formation, in their pathological outcome to the formation of neurotic symptoms. I described this battle between ego and id as terminated by a first truce at the beginning of theatency period and breaking out once more with the first approach to puberty, when the distribution of forces inside the individual is upset by quantitative and qualitative changes in the drives. Threatened with anxiety by the drive development, the ego, as it has been formed in childhood, enters into a struggle for survival in which all the available methods of defense are brought into play and strained to the utmost. The results, that is the personality changes which are achieved, vary. Normally, the organisation of ego and superego alter sufficiently to accommodate the new, mature forms of sexuality. In less favourable instances a rigid, immature ego succeeds in inhibiting ordistorting sexual maturity; in some cases the id impulses succeed in creating utter confusion and chaos in what has been an orderly, socially directed ego during the latency period. I made the point that, more than any other time of life, adolescence with its typical conflicts provides the analyst with instructive pictures of the interplay and sequence of internal danger, anxiety, defense activity, transitory or permanent symptom formation, and mental breakdown.

Interest increased in the postwar years and brought a multitude of contributions, especially from the United States. Fortunately for the student of the subject, Leo a. Spiegel published in 1951 a lengthy "Review of contributions to a Psychoanalytic Theory of Adolescence." Although his attempt to construct an integrated theory out of often widely divergent parts could hardly be successful, the paper serves a most useful purpose by abstracting, reviewing and classifying the material. He groped the publications under headings such as:

"Classification of Phenomenology" (Bernfeld, hartmann,

Kris, and Loewenstein, Wittels)

"Object Relations" (Bernfeld, Buxbaum, H. Deutsch, Erikson, fenihel, A. Freud, W. Hoffer, Jones, A. Katan-Angel, Landauer)

"Defense Mechanisms" (Bernfeld, H. Deutsch, Fenichel, A. Freud, Greenacre, E. Kris)

"Creativity" (Bernfeld, A. Freud)

"Sexual Activity" (Balint, Bernfeld, Buxbaum, H. deutsch, Federn, Ferneczi, S. Freud, Lampl-de Groot)

"Aspects of Ego Functioning" (Fenichel, a. Feud, Harnik, Hoffer, Landauer)

"Treatment" (Aichhorn, K.R. Eissler, A. Freud, Gitelson, A. Katan, M. Klein, Landauer, A. Reich).

A detailed bibliography attached to the review contained altogether forty-one papers by thirty-fur authors, covering apparently every theoretical, clinical, and technical aspect of the subject.

But in spite of this impressive list of contributors and contributions the dissatisfaction with our knowledge of the field remained unaltered, nor did our own, or the parents', confidence in our analytic skill with adolescent patients increase. There was now much published evidence to the contrary; nevertheless, adolescence remained, as it had been before, a stepchild in psychoanalytic theory.

Difficulties of Fact-Finding Concerning Adolescence

There are, we believe, two different causes, which may, possibly, account for our bewilderment when faced with all he intricacies of the adolescent process.

When, in our capacity as analysts, we investigate mental states, we rely, basically, on two methods: either on the analysis of individuals in whom that particular state of mind is in action at the moment, or on the reconstruction of that state in analytic treatment undertaken at a later date, the results of these two procedures, used either singly, or in combination with each other, have taught us all that we, as analysts, know about the developmental stages of the human mind. (It may be worth while to remind the reader in this connection that our knowledge of the mental processes of infancy has been derived from reconstructions in the analyses of adults and was merely confirmed and enlarged on later by analyses or observations carried out in childhood.)

It happens that these two procedures, which have served us well for all other periods of life, prove less satisfactory and less productive of results when applied to adolescents.

Reconstruction of adolescence in adult analysis.—As regard reconstruction, I am impressed how seldom n the treatment of my adult cases I succeed in reviving their adolescent experiences in full force. By that I do not mean to imply that adult patients have an amnesia for their adolescence which resembles in extent or in depth the amnesia for their early childhood. On the contrary, the memories of the events of the adolescent period are, normally, retained in consciousness and told to the analyst without apparent difficulty. Masturbation in preadolescence and adolescent, the first moves toward intercourse, etc, may even play a dominant part in the patients conscious memories and, as we know well, are made use of to overlay and hide the repressed masturbation conflicts and the buried sexual activities of early childhood. Further, in the analyses of sexually inhibited men who deplore the loss of erective potency, it is fairly easy to recover the memories of the bodily practices carried out in adolescence—frequently very crude and cruel ones—which served them at that time to prevent erections, or to suppress them as soon as they occurred. On the other hand, these memories contain no more than the bare facts, happenings and actions, divorced from the affects which accompanied them at the time. What we fail to recover, as a rule, is the atmosphere in which the adolescent lives, his anxieties, the height of elation o depth of despair, the quickly rising enthusiasms, the utter hopelessness, the burning—or at other times sterile—intellectual and philosophical preoccupations, the yearning for freedom, the sense of loneliness, the feeling of oppression by the parents, the impotent rages or active hates directed against the adult world, the erotic crushes—whether home-sexually or heterosexually directed—the suicidal fantasies, etc. These are elusive mood swings, difficult to revive which, unlike the affective states of infancy and early childhood, seem disinclined to re-emerge and be relived in connection with the person of the analyst.

If this impression, which I gathered from my own cases, should be confirmed by other analysts of adults, such a failure—or partial failure—to reconstruct adolescence might account for some of the gaps in our appraisal of the mental processes during this period.

Analysis during adolescence.—When discussing in his "Review" the contributions dealing with the analytic therapy of adolescents, Spiegel 91951) deployed what seemed to him an undue pessimism on the part of some authors. He pointed to the need of adapting the analytic technique to the adolescent Patients' particular situation and expressed surprise at the absence of explicit discussions of an introductory phase "analogous to the one used with children and delinquents."

Actually, since 1951, some further papers on the subject of technique appeared in print. Two of the dealt with the opening phase,(Fraiberg, 1955; Noshpitz, 1957), a third with the terminal one (Adatto, 1958).

While these authors brought material to highlight the special technical difficulties encountered in the beginning and ending of the therapy, work on adolescents done in our Hampstead child-therapy clinic emphasized special technical difficulties met with in the very center of it, i.e., at the critical moment when preadolescence gives way to adolescence proper, when the revolt against the parents is anticipated in the transference and tends to lead to a break with the analyst, i.e., to abrupt and undesirable termination of treatment from the patient's side.

Thus, according to experience, special difficulties are encountered in he beginning, in the middle, and in connection with the end of treatment. Put in other words, this can only mean that the analytic treatment of adolescents is a hazardous venture from beginning to end, a venture in which the analyst has to meet resistances of unusual strength and variety. This is borne out by the comparison of adolescent with adult cases. In adult analysis we are used to handle the difficult technical situations with certain hysterical patients who cannot bear frustration in the transference and try to force the analyst to enact with them their revived low and hate feelings in an actual personal relationship. We are used to guard against the obsessional patients' technique of isolating words from affect and of tempting us to interpret unconscious content while it is divorced from its emotional cathexis. We attempt to deal with the narcissistic withdrawal of the borderline schizophrenics, the projections of the paranoid patients who turn their analyst into the persecuting enemy, the destructive hopelessness of the depressed who claim disbelief in any positive outcome of the analytic effort; the acting out tendencies and the lack of insight of the delinquent or psychopathic characters. But in the disturbances named above, we meet either the one or the other of these technical difficulties, and we can adapt the analytic technique to the resistance which is specific for the type of mental disorder. Not so in adolescence where the patient may change rapidly from one of these emotional positions to the next, exhibit them all simultaneously, or in quick succession, leaving the analyst little time and scope to marshal his forces and change his handling of the case according to the changing need.

Obstacles in the libido economy. Comparison with the states of mourning and unhappy love.—experience has taught us to take a serious view of such major and repeated in adequacies of the analytic technique. Hey cannot be explained away by individual characteristics of the patients under treatment nor by any accidental or environmental factors which run counter to it. Nor can they be overcome simply by increased effort, skill and tact on the part of the analyst. They have to be taken as indications that something in the inner structure of the disturbances themselves differs markedly from the pattern of those illnesses for which the analytic technique has been devised originally and to which it is most frequently applied (Eissler, (1950). We have to gain insight into these divergences of pathology before we are in a position to revise our technique. Where the analyses of children, of delinquents and of certain borderline states are concerned, this has happened already. What the analytic technique had to provide for in these cases were the immaturity and weakness of the patients' ego; their lower threshold for the toleration of frustration; and the lesser importance of verbalization with increased importance of action (acting out) for their mental economy. It remains to be pointed out what corresponding factors are characteristic for the adolescent disorders, i.e., to what specific inner situation of the patients our technique has to be adjusted to make adolescents more amenable to analytic treatment.

So far as I am concerned, I am impressed by a similarity between the responses of these young patients and those of two other well known types of mental upset, namely the reactions to treatment during unhappy love affairs and during periods of mourning. In both these latter states there is much mental suffering and, as a rule, the urgent wish to be helped; in spite of this, neither state answers well to analytic therapy. Our theoretical explanation of this comparative intractability is the following; being in love as well as mourning are emotional states in which the individual's libido is engaged fully in relation to a real love object of the present, or of the most recent pat, the mental pain being caused by the difficult task to withdraw cathexis and give up a position which holds out no further hop for return of love, that is, for satisfaction. While the individual is engaged in this struggle, insufficient libido is available to cathect the person of the analyst, or to flow back regressively and reinvest former objects and positions. Consequently, neither the transference events nor the past become meaningful enough to yield material for interpretation. The immediate object (of love, or of morning) has to be given up before analytic therapy can become effective.

To my mind the libidinal position of the adolescent has much in common with the two states described above. The adolescent too is engaged in an emotional struggle. And moreover in one of extreme urgent and immediacy, his libido is on the point of detaching itself from the parents and of cathecting new objects. Some mourning for the objects of the past is inevitable; so are the "crushes," i.e., the happy or unhappy love affairs with adults outside the family, or with other adolescents, whether of the opposite or the same sex; so is, further, a certain amount of narcissistic withdrawal which bridges the gap urging periods when no external object is cathected. Whatever the libidinal solution at a given moment may be, it will always be a preoccupation with the present time and, as described above, with little or no libido left available for investment either in the past or in the analyst.

If this supposition as to the libido distribution in the adolescent personality can be accepted as a correct statement, it can serve to explain some of our young patients' behaviour in treatment, such as: their reluctance to cooperate; their lack of involvement in the therapy or in the relationships to the analyst; their battles for the reduction of weekly sessions; their unpunctuality; their missing of treatment sessions for the sake of outside activities; their sudden breaking off of treatment altogether. We learn here by contrast how much the continuity of the average adult analysis owes to the mere fact of the analyst being a highly cathected object, quite part from the essential role played by the transference in the production of material.

There are, of course, those cases where the analyst himself becomes the new love object of the adolescent, i.e., the object of the "crush," a constellation which will heighten the young patient's keenness to be "treated," But apart from improved attendance and punctuality, this may mean merely that the analyst is brought up against another of the specific difficulties of the analyses of adolescents, namely the urgency of their needs, their intolerance for frustration and their tendency to treat whatever relationship evolves as a vehicle for wish fulfillment and not as a source of insight and enlightenment.

Under these conditions it is not surprising that besides analytic therapy many alternative forms of treatment for adolescents have been evolved and practiced, such as manipulation of the environment, residential treatment, the setting up of therapeutic communities, etc. Valuable as these experimental approaches are from the practical point of view, they can not, of course, be expected to contribute directly to our theoretical insight into the unconscious contents of the adolescent 's mind, the structure of his typical disturbances, or into the details of the mental mechanisms by which these latter are maintained.

The Adolescent Upset Inevitable

There is, first, the ever recurrent question whether the adolescent upheaval is welcome and beneficial as such whether it is necessary and, more than that, inevitable. On this point, psychoanalytic opinion is decisive and unanimous. The people in the child's family and school, who assess his station eh basis of behaviour, may deplore the adolescent upset which, to them, spells the loss of valuable qualities, of character stability and of social adaptation. As analysts, who assess personalities from the structural point of view, we think otherwise. We know that the character structure of a child at the end of the latency period represents the outcome of long drawn-out conflicts between id and ego forces. The inner balance achieved, although characteristic for each individual and precious to him, is preliminary only and precarious. It does not allow for the quantitative increase in drive activity, nor for the changes of drive quality which are both inseparable from puberty. Consequently, it has to be abandoned to allow adult sexuality to be integrated into the individual's personality. The so-called adolescent upheavals are no more than the external indications that such internal adjustments are in progress.

On the other hand, we all known individual children who as late as the ages of fourteen, fifteen or sixteen show no such outer evidence of inner unrest. They remain, as they a have been during the latency period, "good" children, wrapped up in their family relationships, considerate sons of their mothers, submissive to their fathers, in accord with the atmosphere, ideas and ideals of their childhood background, convenient as this may be, it signifies a delay of normal development and is, as such, a sign to be taken seriously. The first impression conveyed by these cases may be that of a quantitative deficiency of drive endowment, a suspicion which will usually prove unfounded. Analytic exploration reveals that this reluctance to "grow up" is derived not from the id but form the ego and superego aspects of the personality. These are children who have built up excessive defenses against their drive activities and are now crippled by the results, which ct as barriers against the normal maturational processes of phase development. They are, perhaps more than any others, in need of therapeutic help to remove the inner restrictions and clear the path for normal development, however "upsetting" the latter may prove to be.

A second question which we are asked to answer frequently concerns the problem whether the manner in which a given child will react in adolescence can be predicted from the characteristics of his early infantile or latency behaviour. Apart from the more general affirmative answer given by Ernest Jones (1922), only one among the authors named above has made clear and positive assertions in this respect. Siegfried Bernfeld (1923), when discussing his protracted type of male adolescence and its characteristics, established the links between this form of puberty and a specific type of infantile development based on the following three conditions: (a) that the frustration of infantile sex wishes has been shattering for the child's narcissism; (b) that the incestuous fixations to the parents have been of exceptional strength and have been maintained throughout the latency period; (c) that the superego has been established early, has been delineated sharply from the ego, and that the ideals contained init are invested with narcissistic as well as with object libido.

Other and less precise answers to the same question are scattered through the literature. We find the opinion that, in the majority of cases, the manifestations of the adolescent process are not predictable since they depend almost wholly on quantitative relations, i.e., on the strength and suddenness of drive increase, the corresponding increase in anxiety causing all the rest of the upheaval.

I suggested in another place (1936) that adolescence brings about occasionally something in the nature of a spontaneous cure. This happens in children whose pregenital activities and characteristics remained dominant throughout latency until the increase in genital libido produces a welcome decrease in pregenitality. This latter occurrence, on the other hand, can be matched by a corresponding one which produces the opposite effect: where phallic characteristics have remained dominant during latency, the increase in genital libido produces the effect of an exaggerated and threatening aggressive masculinity.

It seems to be generally accepted that a strong fixation to the mother, dating not only from the oedipal but from the preoedipal attachment to her, renders adolescence especially difficult. This latter assertion, on the other hand, has to be correlated with two recent findings of a different nature which we owe to work done in our Hampstead Child-Therapy clinic. One of these findings is derived from the study of orphaned children who were deprived of the relationship to a stable mother figure in their first years. This lack of a mother fixation, far from making adolescence easier, constitutes a real danger to the whole inner coherence of the personality during that period. In these cases adolescence is preceded frequently by a frantic search for a mother image; the internal possession and cathexis of such an image seems to be essential for the ensuing normal process of detaching libido from it for transfer to new objects, i.e., to sexual partners.

The second finding mentioned above is derived from the analyses of adolescent twins, in one case children whose twin relationship in infancy had been observed and recorded in minute detail (Burlingham, 1952). In their treatments it transpired that the "adolescent revolt" against the love objects of infancy demands the breaking of the tie to the twin in no lesser degree than the breaking of the tie to the mother. Since this libidinal (narcissistic as well as object-directed) cathexis of the twin is rooted in the same deep layer of the personality as the early attachment to the mother, its withdrawal is accompanied by an equal amount of structural upheaval, emotional upset, and result symptom formation. Where, on the other hand, the twin relationship survives the adolescent phase, we may expect to see a delay in the onset of maturity or a restrictive hardening of the character of the latency period similar to the instances mentioned above in which the childhood love for the parents withstands the onslaught of the adolescent phase.

To return to the initial question: it seems that we are able to foretell the adolescent reactions in certain specific and typical constellations but certainly not for all the individual variations of infantile personality structure. Our insight in to typical developments will increase with the number of adolescents who undergo analysis.

This leaves us with a third problem which, to my mind, out weight the preceding ones so far as clinical and theoretical significance are concerned. I refer to the difficulty in adolescent cases to draw the line between normality and pathology. As described above, adolescence constitutes by definition an interruption of peaceful growth which resembles in appearance a variety of other emotional upsets and structural upheavals. (Adolescence, of course, is not the only time in life when alterations of a physiological nature cause disturbances of mental equilibrium. The same happens in later years in the climacterium; and recently, Grete L. Bibring (1959) has given a convincing description of similar damager to the equilibrium of mental forces during pregnancy.) The adolescent manifestations come close to symptom formation of the neurotic, psychotic or dissocial order and merge almost imperceptibly into borderline states, initial, frustrated or fully fledged forms o almost all the mental illnesses. Consequently, the differential diagnosis between the adolescent upsets and true pathology becomes a difficult task.

For the discussion of this diagnostic problem leave most other authors in the field to speak for themselves and summarise my own impressions based on past and present clinical experience.

In 1936, when I approached the same subject from the aspect of the defenses, I was concerned with the similarity between the adolescent and other emotional disturbances rather than with the differences between them. I described that adolescent upset take on the appearance of a neurosis if the initial, pathogenic danger situation is located in the superego with the resulting anxiety being felt as guilt; that they resemble psychotic disturbances if the danger lies in the increased power of the id itself. Which threatens the ego in its existence or integrity. Whether such as adolescent individual impresses us, then, as obsessional, phobic, hysterical, ascetic, schizoid, paranoid, suicidal, etc., will depend on the one hand on the quality and quantity of the id contents which beset the ego, on the other hand on the selection of defense mechanisms which the latter employs. Since, in adolescence impulses from all pregenital phases rise to the surface and defense mechanisms from all levels or crudity or complexity come into use, the pathological results—although identical in structure—are more varied and less stabilized than at other times of life.

Today it seems to me that this structural description needs to be amplified, not in the direction of the similarity of the adolescent to other disorders but in that of their specific nature. There is in their etiology at least one additional element which may be regarded as exclusive to this period and characteristic for it; namely that the danger is felt to be located not only in the id impulses and fantasies but in the very existence of the love objects of the individual's oedipal and preoedipal past. The libidinal cathexis to them has been carried forward from the infantile phases, merely toned down or inhibited in aim during latency. Therefore the reawakened pregenital urges, or—worse still—the newly acquired genital ones, are in danger of making contact with them, lending a new and threatening reality to fantasies which had seemed extinct but are, in fact, merely under repression. (An important clinical instance of this can be found in adolescent girls with anorexia nervosa. Here the infantile fantasies of oral impregnation receive added impetus from the new real possibilities of motherhood opened up by genital development. Consequently, the phobic measures adopted against the intake of food on the one hand and identification with the mother on the other hand are overemphasized to a degree which my lead to starvation.) The anxieties which arise on these rounds are directed toward eliminating the infantile objects, i.e., toward breaking the tie with them. Anny katan (1937) has discussed this type of defense, which aims above all at changing the persons and the scene of conflict, under the term of "removal." Such an attempt may succeed or fail, partially or totally. In any case, I agree with Anny Katan that its outcome will be decisive for the success or failure of the other, more familiar line of defensive measures which are directed against the impulses themselves. A number of illustrations will serve to clarify the meaning of this assumption.

Defense by displacement of libido.—there are many adolescents who deal with the anxiety around by eh attachment to their infantile objects by the simple means of flight. Instead of permitting a process of gradual detachment from the parents to take place, they withdraw their libido from them suddenly and altogether. This leaves them with a passionate longing for partnership which they succeed in transferring to the environment outside the family. Here they adopt varying solutions: Libido may be transferred, more or less unchanged inform, to parent substitutes, provided that these new figures are diametrically opposed in every aspect (Personal, social, cultural) to the original ones. Or the attachment may be made to so-called "leaders'," usually persons in age between the adolescent's and the parents' generation, who represent ideas. Equally frequent are the passionate new ties to contemporaries, either of the same or of the opposite sex (.e., homosexual or heterosexual friendships) and the attachments to adolescent groups (or "gangs"). Whichever of these typical solutions is chosen, the result makes the adolescent feel "free,' and revel in a new precious sense of independence form the parents who are treated, then, within difference bordering on callousness.

Although the direction taken by the libido in these instances is, in itself, on lines of normality, the suddenness of the change, the carefully observed contrast in object selection, and the overemphasis on the new allegiances mark it as defensive. It represents an all too hasty anticipation of normal growth rater than a normal developmental process. It makes little further difference to the emotional situation whether the libidinal flight is followed by actual flight, i.e., whether the adolescent also "removes" himself bodily from his family. If not, he remains in the home in the attitude of a boarder, usually a very inconsiderate one so far as the older and younger family members are concerned.

On the other hand, the withdrawal of cathexis from the parents has most decisive consequences for the rest of the defensive processes. Once the infantile objects are stripped of their importance, the pregenital and genital impulses cease to be threatening to the same degree. Consequently, guilt and anxiety decrease and the ego becomes more tolerant. Formerly repressed sexual and aggressive wishes rise to the surface and are acted on, the actions being taken outside the family in the wider environment. Whether this acting out will be on harmless, or idealistic, ordissocial, or even criminal lines will depend essentially on the new objects to which the adolescent has attached himself. Usually the ideals of the leader, of the adolescent group, or of the gang, are taken over wholeheartedly and without criticism.

Adolescents of this type may be sent for treatment after their actions have brought them into conflict with their schools, their employers, or the law. As far as psychoanalytic therapy is concerned, they seem to offer little chance for the therapeutic alliance between analyst and patient without which the analytic technique cannot proceed. Any relationship to the analyst and, above all, the transference to him would revive the infantile attachments which have been discarded; therefore the adolescent remains unresponsive. Also, the escape from these attachments has suspended the feeling of internal conflict, at least temporarily; consequently, eh adolescent does not feel in need of psychological help. A. Aichhorn had these points in mind when he maintained that adolescents of this is social and criminal type needed a long period of preparation and inner rearrangement before they could become amenable to analytic treatment. He maintained that the latter would be successful only if, during this preparation in a residential setting, the adolescent made a new transference of object love, reawakened his infantile attachments, internlised his conflicts once more, in short, became neurotic.

To try and analyse adolescent in his phase of successful detachment from the past seems to be adventure doomed to failure.

Defense by reversal of affect.—A second typical reaction to the same danger situation is, although less conspicuous outwardly, more ominous in nature inwardly.

Instead of displacing libido from the parents—or, more likely, after failing to do so—the adolescent ego may defend itself by turning the emotions felt toward them into their opposites.

This changes love into hate, dependence into revolt, respect and admiration into contempt and derision. On the basis of such reversal of affect the adolescent remains himself to be "free' but, unluckily for his peace of mind and sense of conflict, this conviction does not reach further than the conscious surface layer of his mind. For all deeper intents and purposes he remains as securely tied to the parental figures as he has been before; acting out remains within the family; and any alterations achieved by the defense turn out to his disadvantage. There are no positive pleasures to be derived from the reversed relationships, only suffering, felt as well as inflicted. There is no room for independence of action, or of growth; compulsive opposition to the parents proves as crippling in this respect as compulsive obedience to them can prove to be. (Ferenczi has pointed to this effect of "compulsive disobedience" many years ago.) Since anxiety and guilt remain undiminished, constant reinforcement of defense is necessary. This is provided in the first place by two methods: denial (of positive feeling) and reactions formations (churlish, unsympathetic, contemptuous attitudes). The behavioural picture that emerges at this stage is that of an uncooperative and hostile adolescent.

Further pathological developments of this state of affairs are worth watching. The hostility and aggressiveness, which serve as a defense against object love in the beginning, soon become intolerable to the ego, are felt as threats, and are warded off in their own right. This may happen by means of projection; in that case the aggression is ascribed to the parents who, consequently, become the adolescent's main oppressors and persecutors. In the clinical picture this appears first as the adolescent's suspiciousness and, when the projections increase, as paranoid behaviour.

Conversely, the full hostility and aggression may be turned away from the objects and employed inwardly against the self. In these cases, the adolescents display intense depression, tendencies of self-abasement and self-injury, and develop, or ever carry out, suicidal wishes.

During all stages of this process, personal suffering is great and the desire to be helped intense. This, in itself, is no guarantee that the adolescent in question will submit to analytic therapy. He will certainly not do so if treatment is urged and initiated by the parents. Whenever this happens, he will consider analysis as their tool, extend his hostility or his suspicions to include the person of the analyst and refuse cooperation. The chances are better if the adolescent himself decides to seek help and turns to analysis, as it were in opposition to the parents' wishes. Even so, the alliance with the analyst may not be of long duration. As soon as a true transference develops and the positive infantile fantasies come into consciousness, the same reversal of affect tends to be repeated in the analytic setting. Rather than relive the whole turmoil of feelings with the analyst, many adolescent patients run away. They escape from their positive feelings, although it appears to the analyst that they break off treatment in an overwhelmingly strong negative transference.

Defense by withdrawal of libido to the self.—to proceed in the direction of increasing pathology:

Withdrawal of libido from the parents, as it has been described above, does not, in itself, decide about its further use, or fate. If anxieties and inhibitions block the way toward new objects outside the family, the libido remains within the self. There, it may be employed to cathect the ego and superego, thereby inflating them. Clinically this means that ideas of grandeur will appear, fantasies of unlimited power over other human beings, or of major achievement and championship in one or more fields. Or, the suffering and persecuted ego of the adolescent may assume Christ-like proportions with corresponding fantasies of saving the world.

On the other hand, the cathexis may become attached to the adolescent's body only and give rise there to the hypochondriacal sensations and feeling of psychotic illness.

In either case analytic therapy is indicated as well as urgent. Treatment will dispel the appearance of severe abnormality if it reopens a path for the libido, either to flow backwards and recathect the original infantile objects, or to flow forward, in the direction described above, to cathect less frightening substitutes in the environment.

What taxes the analyst's technical skill in these cases is the withdrawn state of the patient, i.e., the problem of establishing an initial relationship and transference. One this is accomplished, the return from narcissistic withdrawal to object cathexis will relieve the patient, at least temporarily.

I believe, there are many cases where the analyst would be wise to be content with this partial success without urging further treatment. A further, and deeper, involvement in the transference may well arouse all the anxieties described above and, again, lead to abrupt termination of the analysis due to the adolescent's flight reaction.

Defense by regression.—The greater the anxiety aroused by the object ties, the more elementary and primitive is the defense activity employed by the adolescent ego to escape them. Thus, at the extreme height of anxiety, the relations with the object world may be reduced to the emotional state known as "primary identification' with the objects. This solution with which we are familiar from psychotic illnesses implies regressive changes in all parts o the personality, i.e., in the ego organisation as well as in eh libido. The ego boundaries are widened to embrace parts of the object together with the self. This creates in the adolescent surprising changes of qualities, attitudes and even outward appearance. His allegiance to persons outside himself betrays itself in these alterations of his own personality (.e., his identifications) rather than in an outflow of libido. Projections, together with these identifications, dominate the scene and create a give-and-take between the self and object which has repercussions on important ego functions. For example, the distinction between the external and internal world (i.e., reality testing) becomes temporarily negligible, a lapse in ego functioning which manifests itself in the clinical picture as a state of confusion.

Regression of this kind may bring transitory relief to the ego by emptying the oedipal (and many of the preoedipal) fantasies of their libidinal cathexis. But this lessening of anxiety will not be long lived. Another and deeper anxiety will soon take its place which I have characterized on a former occasion (1951) as the fear of emotional surrender, with the accompanying fear of loss of identity.

Defense Against Impulses

Where the defenses against the oedipal and preoedipal

object ties fail to achieve their aim, clinical pictures emerge which come nearest to the borderline toward psychotic illness.

The "ascetic" adolescent.—One of these, the "ascetic" adolescent, I have described before as fight in all his impulses, preoedipal and oedipal, sexual and aggressive, extending the defense even to the fulfillment of the physiological needs for food, sleep, and body comfort. This, to me, seems the characteristic reaction of an ego, driven by the blind fear of over whelming id quantities, an anxiety which leaves no room for the finer distinctions between vital or merely pleasant satisfactions, the healthy or the morbid, the morally permitted or forbidden pleasures. Total war is waged against the pursuit of pleasure as such. Accordingly, most of the normal processes of instinct and need satisfaction are interfered with an become paralyzed. According to clinical observation, adolescent ascerticism is, with luck, a transitory phenomenon. For the analytic observerit provides precious proof of he power of defense, i.e., of the extent to which the normal, healthy drive derivative are open to crippling interference by the ego.

On the whole, analytic treatment of the ascetic typedoes not presnt as many technical difficulties as one would expect. Perhaps, in these individuals, defense against the impulses is so massive, that they can permit themselves some object relationship to the analyst and, thus, enter into transference.

The "uncompromising" adolescent.—Another, equally abnormal adolescent, is described best as the "uncompromising" type. The term, in this instance, does refer to more than the well-known conscious, unrelenting position adopted by many young people who stand up for their ideas, refuse to make concessions to the more practical and reality-adapted attitudes of their elders, and take pride in their moral or aesthetic principles. "Compromise," with these adolescents, includes processes which are as essential for life as, for example, the cooperation between impulses, the blending of opposite strivings, the mitigation of id strivings by interference from the side of the ego. One adolescent whom I observed in analysis did his utmost, in pursuit of this impossible aim, to prevent any interference of his mind with his body, of his activity with his passivity, his loves wit his hates, his realities with his fantiatises, the external demands with his internal ones, in short, of hi ego with his id.

IN treatment this defense was represented as a strong resistance against and "cure," the idea of which he despised in spite of intense suffering. He understood correctly that mental health is based in the last resort on harmony, i.e., on the every compromise formations which he was trying to avoid.

The Concept of Normality in Adolescence

Where adolescence is concerned, it seems easier to describe its pathologicalmanifestatios thanthe normal processes. Nevertheless, there are in the above exposition at least two pronouncements which there are in the above exposition at lesttwo pronouncements which may prove useful for theconcept: (1) that adolescence is by its nature an intrruption of peaceful growth, and (2) that the upholding of a steady eqilibrium during the adolescent process is in itself abnormal. Once we accept for adolescence disharmony within the psychic sructure as our basic fact, understanding becomes easier. We beginto see the upsetting battles which are raging between id and ego as beneficient attempts to restore peace and harmony. The defensive methods which are employed either against the impulses, or against the object cathexis, bein to apper legitimate and normal. If they produce pathological results, this happens not because of any malignancy in their nature, but because they are overused, over stressed, orused in isolation. Actually, each of the abnormal types of adolescent development, as it is described above, represnts also a potentially useful way of regaining mental stability, normal if combined with other defenses, and if used in moderation.

To explains last statement in greater detail: I take it that it is normal for an adolescent to behave for a considerable length of time in an inconsistent and unpredictable manner; to fight his impulses and to accept them; to ward them off successfully and to be overturn by them; to love his parents and to hate them; to revolt against them and to be dependent on them; to be deeply ashamed to acknowledge his mother before others and, unexpectedly, to desire heart-to-hart talks with her; to thrive on imitation of and identification with others while searching unceasingly for his own identity; to be more idealistic, artistic, generous and unselfish than he will ever be again, but also the opposite; self-centered, egoistic, calculating. Such fluctuations between extreme opposites would be deemed highly abnormal at any other time of life. At this time they may signify no more than that an adult structure of personality takes a long time to emerge, that there go of the individual in question does not cease to experiment and is in no hurry to close down on possibilities. If the temporary solutions seem abnormal to the onlooker, they are less so, nevertheless, than the hasty decisions made in other cases for one-sided suppression, or revolt, or flight, or withdrawal, or regression, or asceticism, which are responsible for the truly pathological developments described above. While an adolescent remains inconsistent and unpredictable in his behaviour, he may suffer, but he does not seem to me to be in need of treatment.

REFERENCES

Bibring, G.L. (1959), Some considerations of the psychological processes in Pregnancy. *The Psychoanalytic Study of the Child,* XIV, New York: International Universities Press.

Eissler, K.R. (1950), Ego-Psychological Implications of the Psychoanalytic treatment of Delinquents, *The Psychoanalytic Study of the Child,* V. New York: International Universities Press.

Geleerd, E.r, (1957), Some Aspects of Psychoanalytic Technique in Adolescents. *The Psychoanalytic Study of the child,* xii, New York:

International Universities Press.

Harnik, J. (1923), The Various developments Undergone by Narcissism in Men and Mewn, *Int. J. Psa.,* V. 1924.

Katan, M. (1950), Structural Aspects of a Case of Schizophrenia. *The Psychoanalytic Study of the Child,* V. New York International Universities Press.

Maenchen, A. (19460, A Case of Superego Disintegration. *The psychoanalytic Study of the Child,* II, New York: International Universities Press.

11

CHILD GUIDANCE CLINIC AS A CENTRE OF PROPHYLAXIS AND ENLIGHTENMENT

Child Analysis before the child Guidance Era begin with my personal recollections about the development of these relationships, recollections which make me fele that analysts and child analysts owe the child guidance clinics a debt of gratitude. Before such clinics existed, we analysts felt that we were in possession of much solid analytic knowledge potentially useful for the upbringing of children, their teaching, and the treatment of their problems. But there were no recognized channels of communication, neither with parents nor with the professional workers within he children's services. All that the analysts could rely on in this respect were chance contacts: parents while or after undergoing a therapeutic analysis themselves could not help but apply some of their newly acquired insight to the handling of their children; teachers who for personal reasons, underwent analysis were expected to carry back to the teaching profession some analytic findings concerning child development It was a thrilling occasion when in Vienna (approximately in 1930) an inspector of nursery schools invited us to conduct a seminar for a small group of nursery school teachers to discuss their problem children; or when an inspector of municipal play centres invited me to give a series of introductory lectures to his workers. Actually, one of the first schools to make systematic use of analytic child psychology was the Walden School in New York, a fact which Dr. S. Ferenczi learned on one of his early visits to the States and reported on most enthusiastically after his return. But these hopeful sins of possible cooperation between analysis and work with children remained few and far between.

There was another side to the same picture which was no less frustrating, We knew that there was a large amount of child material near at hand from which we, as child analysts, could have profited greatly. There were all the initial states of mental disturbance, as they show up in childhood, the many transitional stages between health and illness, the common developmental difficulties of all young children, their relationships to the more circumscribed syndromes of the infantile neuroses—all this was potentially available in our environment and, still, we were denied access to it. The children brought to us in private practice or within the hospital services were those where mental illness had passed the initial phase of behaviour disorder or symptom formation and had become ingrained in the personality in a massive way, causing severe breakdowns or permanent deviations of character. Again, there was little or nothing that we could do to alter the position.

But it is precisely this situation which has changed with the advent of the child guidance clinics. Since they have come into being they have opened up a two-way traffic between parents and professional workers with children on the one hand and child analysis or child psychiatry on the other hand. It is the child guidance clinics which act at present as mediators between the two fields and serve as clearing house between them: they receive material from the public to pass on to the child analysts and child psychiatrists, and conversely they receive information from child psychiatry (or analysis) to hand out to the public. This is an important and formidable assignment, and it seems worth while to me to investigate how far the child guidance clinics of our knowledge have been able and equipped to fulfill this double task.

The Child Guidance Clinic as a Clearing House

As regards the first side of the assignment, the collection of available case material, there is no doubt tat the existing clinics are eminently successful. Probably on the basis of their status as public institutions they have done what they private child analyst had been unable to achieve: namely, to gain the confidence of the public. (It was August Aichhorn who remarked first on this difference between private and public work with parents. When praised for the hold he hand on parents and the amount of information he gathered from them, he would answer: "don't forget what I represent for the parents. As official advisor, appointed by the authorities, I am backed up by the whole Municipality.") As director of a child guidance clinic, or when talking to one of directors, you will find that collecting the material presents no difficulty. If anything, trouble arises in the opposite direction; there is too much material. Clinics are swamped with the cases sent in to them by the parents themselves, by teachers, youth workers, general practitioners and pediatricians, almoners of hospitals, probation officers from the courts, child welfare organisers, etc. There is everything from the most common difficulties of management with which parents used to deal on their own (such as the early sleeping, feeding, and toileting troubles), to the severe retardations, borderline and autistic and atypical states. There is, indeed, more than the diagnostic skill o the child psychiatrist or child analyst can cope with at present. The bewildering variety of pathological, or seemingly pathological manifestations, seems to call for new diagnostic categories based not on symptomatology but on developmental considerations. But, however this may be, the case material is there to be looked at by the analyst, studied, sorted out for therapy, understood, and, perhaps, in the future, to be newly classified.

There is, as said before, no lack of confidence in the clinic on the part of the public; again, if anything, thee is too much. With many people, the concept of treatment in a clinic has acquired a magical connotation. Little difference is made between children who suffer from environmental neglect, unsuitable parents, tragic bereavements, and those cases where something has gone wrong in the development and personality structure for internal reasons. In both instances to know that a child is in treatment seems to spell safety to parents, welfare agencies or public authorities, as if ever non-organic disturbance were automatically suitable for psychological treatment, and as if treatment guaranteed a cure in all cases. Paradoxically enough, it will soon be necessary on the part of the clinics to disabuse the public, i.e., to induce people to lower their expectations of therapy to a level more justified by the conditions and limitations imposed by reality.

The Application of Analytic Knowledge

The clinics have been less successful where the receiving and application of analytic knowledge is concerned. At lest, there is in this respect a wide gap between the "analytic" clinics as they exist in some of the big European and American centres, usually in close collaboration with a local psychoanalytic society and training institute, and the ordinary, although frequently "analytically oriented" child guidance clinics as they exist in larger or smaller numbers allover the countries. The analytic clinic usually has a fully trained analyst as psychiatrist-in-charge with several younger staff members in analytic training. The analytically oriented clinic may base its whole connection with psychoanalysis on some contact with analysis which a senior staff member has had at one period of his career. What has been brought back from such contacts are often no more than single pieces of analytic findings which bear the hallmark of a certain phase of analytic thinking. I name as such the following; emphasson the pathogenic role of repression of infantile sexuality (resulting in therapeutic emphasis on sex enlightenment, dealing with guilt arising from oedipal fantasies, masturbation conflicts, etc.): emphasis on the pathogenic repression of aggression (with therapeutic emphasis laid on the discharge of the child's aggressive and destructive tendencies); emphasis on the mother-infant unity in the earliest years (treatment being directed toward the aftereffects of traumatic separations); emphasis on the pathogenic role of the mother's abnormal personality, emotional withdrawnness, severely neurotic or psychotic traits (with consequent inclusion of the mother in the child's treatment). Not that these findings do not represent all-important parts of the analytic knowledge of child development as it stands at present, but they have to be pieced together and fitted into the framework of analytic child psychology before their clinical application can be wholly useful. When they are used in isolation, the diagnostic as well as the therapeutic issues become distorted, a frequent occurrence for which psychoanalysis has had to take much blame. Perhaps in future our distinction between analytic child guidance clinics and others should be based less on the therapeutic time available for each patent (one weekly versus daily treatment) but on the fact whether the whole analytic knowledge of children is available for application or merely selected portions of it.

We are left with the question where to place the responsibility for this, to the analyst, unsatisfactory state of affairs and where to look for remedies. The fault may lie mainly with us analysts who are notoriously bad at public relations and known to publicize findings all too often in unsuitable language or with exaggerated emphasis on the most recent discoveries. On the other hand, the fault may lie with the clinics where people fail to sift, weight, and integrate data before applying hem to therapy and guidance work. The necessity for such discrimination in this field is not so different, after all, from the corresponding one in the medical field where knowledge of new findings, new methods or new drugs has to be pursued constantly and integrated with former knowledge before application.

The Distribution of Analytic Knowledge

For discussing the distribution of knowledge I turn once more to the experience of those clinics where a considerable fund of reliable and comprehensive analytic data is available for use. The points of contact with the public open to them are of various kinds.

Diagnostic Assessment

Thee are, for one, the constant demands for the assessment of children, as they are made on all child guidance clinics for purposes of the juvenile courts, or placement for adoption, with foster parents, in residential institutions, for selection of school, etc. Where the assessment is made not in terms of symptomatology but from the wider point of view of personality structure and development (as mentioned above), and where the language is non-technical, more than the immediate practical purpose can be served. Here is a legitimate channel of communication which can be used to acquaint the referring agencies with the clinical (and theoretical) aspects on which the clinic's understanding of the child is based. In time these case histories will illustrate many of the main issues of analytic child psychology, such as the interdependence of environmental and internal factors, of body and mind, of emotional and intellectual capacities; the overding importance of drive activity and its position in the personality structure; the role of anxiety and of defense activity leading to inhibition of function or to symptom formation; the developmental aspects of social or dissocial development, etc. (In the East London child-Guidance Clinic Dr. Augusta Donnard has carried out valuable work of this kind of or a number of years, She has done an effective teaching job with the school and welfare authorities in her area through answering their requests for diagnostic assessment by means of elaborate and enlightening developmental pictures of the children referred to her.)

Instruction of Professional Workers

A second and more direct channel of communication is opened up by the constant demands for instruction which arrive from the professional workers in the field, such as pediatricians, nurses, or teachers. It is true that many of us feel that the teaching role of the clinic is a very temporary one with regard to these. By rights, instruction in child development should be an integral part of the medical or nursing courses as well as of the teachers' training colleges. But while developments in these institutions have to be waited for, there is a gap to fill, and it may be worth while to discuss how to do this constructively, avoiding past mistakes. (The next speaker in the Symposium, Dr. Grete Bibring, will report on the manner in which the nalytic theory of child development is taught today already to pediatricians and other residents in the Beth Israel hospital, Boston, Mass, Another participant in the Convention. Dr. Anny Katan, might tell us about the way in which such knowledge is integrated with the medical course form the first year onward in Western Reserve University, Cleveland, Ohio.)

So long as the general knowledge of child development has to be introduced as an afterthought to the majority of the members of the medical, nursing, and teaching professions, the success or failure of such ventures will depend on three factors: 910 the selection of the right teaching method: (2) the selection of the right facts; (3) some over-all considerations which provide a framework.

To take the lat point first: when working with doctors, nurses, or teachers, it is essential to remember that not all the psychological data which we can offer are suitable for immediate application, since the condition under which they work are different from ours. It is only a short time ago that teachers learned that the child's intellect does not function divorced from his feelings and that, therefore, some connection with their pupils' emotional life is indispensable for teaching. For the medical world it is a similarly recent discovery that the child's body is not divorced from his mind and that, therefore, physical and mental matters are interrelated sufficiently to influence and interfere with each other. But schools and hospitals as institutions have antedated all these new findings. The application of the latter, therefore, implies a whole host of changes which range from the free entry of parents to schools and hospitals to the rota of changing nursing personnel and the very arrangements of school room or ward, changes which especially for nurses and doctors may mean a more or less complete upheaval in the established conditions of their work.

As regards teaching methods, many of us agree that formal courses in child development remain effective before links have been established between the data offered in them and those contained in formal professional training on physical matters. On the other hand, openings are provided by those patients or pupils how remain "problems" in medical practice or school room, i.e., where the physical or intellectual data fail to explain the picture. Teaching "on the case" has been found satisfactory by many analysts for these reasons.

As for the selection of facts for teaching, I emphasize, as before, that few analytic data are beneficial if applied out of context. Schools, for example, may profit greatly from applying the analytic concept of sublimation to the teachers' work. By deflecting libidinal energy from its original sources and harnessing it to school subjects much drudgery in school can be changed into activities pleasurable almost as play. On the other hand, progressive schools will come to grief if they base their schedules too consistently on this single piece of analytic knowledge. Psychoanalytic theory also has much to say about the essential differences between play and work, the first being governed by the pleasure principle, the latter by the reality principle. It is one of the main tasks of education to lead each individual child from the earlier to the later mode of functioning. Only the integration of the knowledge of sublimation with the ore intricate knowledge of mental functioning on the various stages of development will be of real benefit to the psychological orientation of the modern schools.

Pediatricians might be similarly misled if they meet the present psychoanalytic campaign against hospitalization of small children unconnected with other facts. It is true that a first early separation form home for purposes of medical or surgical intervention may prove traumatic to the individual child and that the psychological damage done by it may go as far as out weighting the physical benefit in health. On the other hand, doctors should not be impressed exclusively with the fact, or the effects, of separation. It is equally necessary for them to learn what part physical illness as such even when undergone at home plays in the mental development of children. Again, it is only the combination and integration of both these lines of inquiry which will lead benefit application.

Work with Parents

This leads us to the third and lat channel of communication with the pubic, the work with parents which will remain the Clinic's domain, whatever the developments.

There is at present the widespread belief, held by all clinic personnel, that work with parents (or in the case of young children with their mothers) is effective only if directed toward changes in their personalities. Consensus on this point has been reached, I believe, on the rebound from earlier and outmoded attitudes toward the parents, attitudes which included much teaching of facts on a purely intellectual basis, advising in an authoritative manner or, at worst, preaching and exhorting. Our analytic discoveries have laid bare the intricate emotional relationships between child and parent, the role of internalization of the parents' personalities, and identification with their qualities, the consequences of love and hate, the subtle or crude forms of seduction the overriding importance of any one of the developmental phases, whether libidinal or aggressive, being acceptable or repulsive to the parent. Once the parent-child relationship is seen in this light, work wit a parent ceases to imply a teacher-pupil attitude and takes on the aspect of a therapeutic relationship.

Most analysts subscribe to these arguments, which are, after all, derived from their work, although with the provide that experience in analytic practice also teachers to be modest in the expectation of personality changes through any but the most intensive therapy.

For my own person, I remain a heretic in some of these respects. I am far from denying the effect on a child from a beneficial change in he parents' attitudes. Still, I cannot help believing that this therapeutic approach to the parents is not more than one among a whole series of possible approaches, all of them serving the same ultimate purposes of beneficial change, prevention, and enlightenment. In what follows I shall try to justify this rather unpopular view of mine by outlining a number of these possibilities as I have learned to see them.

The therapeutic approach

Those instances where we have no hesitation and no reservation in joining the popular demand for therapeutic dealings with the parents. There are the cases, by now known to most clinics, where the mother's and the child's pathology are completely intertwined. We believe that very young children have some primitive pathway of communication with their mothers' unconscious which render them sensitive and vulnerable for the letter's unconscious libidinal or aggressive impulses and fantasies. A young child may merely act out the mother's fantasy in play, or may defend himself against it in some complicated way, or may build a complex pathological structure of his own on the basis of the mother's secret or manifest disturbance. In the most extreme instances the mental disorder of mother and child will be fitted together as if they were pieces in a jug saw puzzle or—to compare it with adult disorders—as in a *folie-adeux*.

In our Hampstead Clinic we have studied a number of such mother-child couples over the years by analyzing both partners simultaneously under different therapists and analysts and coordinating the results. Our aim is to determine the limits of dependence and independence in their disturbances and to assess the therapeutic chances for those other cases where only the child, i.e., the more passive factor in the partnership, is taken into treatment. On the basis of several of these investigations we have concluded that there is some hope in a altering the condition by the child's analysis alone so long as the pathogenic agents from the mother's side remain in the realm of thought and fantasy; but that nothing except simultaneous treatment will alter the position where these mental influences are reinforced by actions of the mother which tie the child to her as a result of pleasurable or painful actual excitations.

The therapeutic approach to the mother is equally vital in those cases where young children show the various signs of deprivation of mothering, though in the mothers presence. This may be caused by a depression of the mother, or period of actual mourning which causes withdrawal of feeling from the object world, including the child. Her incapacity to fulfill the task of mothering may be caused, equally, by the consequences of a pathogenic masculinity complex, etc. Again, nothing except therapy will have a chance of altering the position. But, since the therapeutic results are slow to come, we have seen repeatedly that the change in the mother comes too late for the child on whose handling he deficiency became apparent and that it is the next born infant who benefits from the preventive measure. (Mrs. Kata Levy, London, is responsible for drawing our Clinic's attention to the difference between the mother's disturbance being (a) a direct agent in the child's pathology through identification and excitation; (b) an indirect agent through incapacitating the mother for her task of mothering.)

The Semitherapeutic Approach: Guiding the Parents through the Child's Analysis.—The close interdependence of mother and child t the beginning of life, and the exaggeration of this factor in some abnormal cases, should not make us ignore the fact that children are growing personalities in their own right with a wide range of possible disturbances and conflicts within their structure. But while they need treatment in their own right, their mother will be severely affected by this procedure and will need to be helped and guided through the period. To safeguard the child's future in the family after conclusion of treatment, the parents, and especially the mother, should be allowed or even urgent to accompany the analytic process to some degree. This implies facing their own adult resistances against the repressed remnants of infantile sexually and the cruder forms of infantile aggression, all the more so since the child's impulses are directed against their persons and become increased temporarily during treatment as a rule. The child's treatment acts as a threat to the defenses of the mother, quite apart form the violent feelings of jealousy and competition with the child analyst which are aroused by the young patient's positive attachment to the latter's person. Much human tact plus therapeutic skill on the part of the clinic's representative (whether therapist, psychiatrist, or psychiatric social worker) will be needed to handle these difficulties of the parents and to safeguard treatments.

The therapeutic-Guiding Approach; Dealings with the Mothers of Handicapped Children—the approach is different again where the mothers of abnormal children are concerned, i.e., those who are either born physically or mentally defective (blind, deal spastic, deformed, Mongol, etc.) or have acquired such abnormalities in early life, or undergo some unusual experience or fate (traumatic separations, deprivations, adoption, fostering, mother, or dietary restrictions). There is no reason to expect that mothers are equipped automatically for the specialised as of bringing up such a child in a manner calculated to minimize the handicap. On the contrary, the mother's natural hurt and despair concerning her child's defect, the injury to her pride and pleasure in the child, will all work toward estranging her from the task of mothering thereby increasing the initial damage. There is her ea specially difficult therapeutic task with the mothers which has been tackled in few places. But there is, besides it, also the need for expert advice, based on knowledge of the abnormalities concerned, such as the following: how to ensure with the blind infant that contact with the object world by sight is substituted for through other channels; how to provide motor outlet for the blind toddler, thereby preventing many of the abnormalities following on motor restriction; how to prevent necessary dietary restrictions from being understood by the child as punishments and experienced as intolerable deprivations; how to minimize the effect of inevitable separations from the mother; how to prepare for operations and meet their equally inevitable aftereffects, etc., etc. There are a whole host of emergency situations of this kind in which the normal mother will feel helpless without guidance.

The Guiding approach; Treatment of the Young Child by the Mother.—A number of analytic clinics, besides ours, have devised a procedure recently in which an intuitive and willing mother, free from severe pathology herself, is helped to guide hr own child through the phases of an average developmental disturbance or infantile neutosis. Such work implies close contact with the mother usually through weekly sessions, in which the child's difficulties, his tantrums, his behaviour, his play or whatever other material has been elicited are discussed, assessed, and elucidated for the mother who then bases her next handling of the child on her understanding. Inevitably this implies also much reassurance of the mother, so far as the severity of the pathology, the comparison with other children, the confidence in her own ability as a mother are concerned. I have reported elsewhere personal experience with two cases of this kind, one a case of soiling in a two-and-a-half-ear-old boy, the other an incipient but violent school phobia in a six-year-old girl. In both instances satisfactory results were achieved, n spite of the therapeutic aspects of the work being concentrated on the child alone. That the mother's attitudes during the cooperation have to be met with understanding, patience, and sympathy goes without saying. Also, the mothers suitable for this type of cooperation have to be selected carefully; but there are more than a few who will be able to respond positively to the appeal to their reason, common sense, and good will, who will cope with their own resistances and succeed in the task, helped by the satisfaction that their child need not be handed over to a stranger for therapy.

The Indirect Approach by Way of Public Opinion—It is not only on the basis of the foregoing considerations that I refuse to believe that mothers need to change their personalities before they can change the handling of their child. Assertions of this kind, I find, are not based on facts, above all not on historical facts. Even in the last 30 to 40 years during which was able to observe the methods of upbringing, changes have been remarkably great. Beast feeding, for example, has been adopted, given up, is being adopted again, and may be on the way out once more. Feeding schedules have been extremely strictly observed for two or three decades (with some disastrous results for the later feeding disturbances), have given way to feeding on demand (with other disturbing results), and show some indication at present of returning with more moderation. Toilet training before the war in England began with official hospital approval soon after birth, was delayed in the postwar period until the end of the first year, and is postponed now by man mothers until the second or even third year. To widen the picture we need think only of the question of sexual enlightment now accepted as a must by almost al parents; of the immensely increased freedom given to the child's aggression, even if directly expressed toward the parents; of the altered attitude to corporal punishment as a recognized tool of education, etc. Far-reaching developments of this kind are not due directly to personality changes in the individual mothers but indirectly to a alterations in the social atmosphere to which the mothers cannot help but respond. My contention is, therefore, that in rearing their children mothers are not only guided by instinct and misled by distorting personal influences, but they are to an even larger degree dependent on tradition and public opinion, both of which are open to change.

For the child guidance clinic this opens up the possibility of playing a part in the setting up of new traditions. As described above, they have gained the confidence of the public already where the treatment of problem children is concerned. T should not be difficult for them to establish a similar position of trust with regard to the handling of normal children during the all-important first four or five years of life, very much in the manner in which the well-baby clinics have established their traditions with regard to the healthy infants' physical needs. This will be slow work, supplanting gradually what is left of religious, national, and class traditions, but it will e no less effective so far as influence on the mothers is concerned. Further role for the child guidance clinics. At present all the workers in the children's services, whether in schools, hospitals, courts, clinics, suffer from the effects of specialized training and lack of coordination and integration in the field. It is as rare for a teacher to deal with a sick child as it is for a hospital nurse to handle a healthy one, or a pediatrician to come in contact with the juvenile courts. This limits each specialists' outlook on childhood of which the various tasks show no more than a single aspect.

During at lest the ten years that we have been here, here, we have never been able to meet the demands of the community for child psychiatric services. Since cannot possibly do all the things relating to child psychiatry that need to be done, we probably meet only a small fraction of the need. We know that it is unlikely that we will be able to do so in the foreseeable future. What needs shall we meet? To what extent and how? This is our problem. We hoped to make order out of the choas that existed when the clinic was operated by responding to pressures for service. I was only the realization of the hopelessness of trying to be governed by pressure for services that led to the discarding of that method and taking the reins in our own hands. In this procedure the first thing that had to be decided was how to handle the pressures on the clinic so that we could effectively work in peace.

In this chapter deal only with that phase of the program which deals with the problem of diagnosis and selection for treatment which is vital to the whole plan. There existed a long waiting list for treatment, and the cases that survived the waiting list were usually the most serious and least hopeful ones, requiring long term treatment. The clinic was at the same time loaded down with long, almost interminable cases; hence, the longer the system continued, the less new cases we were able to take on. The most serious cases are the ones with the greatest emotional appeal to the staff so that this also meant they were most likely to be accepted. These cases also were not necessarily the most treatable kind, and we felt that in the long run we were not using our staff time as efficiently was we could by letting the cases select themselves wither by pressure, survival of the waiting list by the continuation of crises, or by playing on our feelings where we felt too guilty to reject distressing cases.

We began with a policy of not accepting more than 10 per cent long-term cases of dubious prognosis. The policy was fine, but seemed unenforcible. Either the cases were misdiagnosed of each staff member presenting a case for admission found special reasons why an exemption should be made in each instance. The reasons were valid, but so was the need to balance the case load of the clinic. We found that we could not get the staff to say no to a parent when there was urgent need for treatment regardless of the treatability of the case or the over-all needs of the community.

We next grappled with this problem and felt that it was impossible to teach staff members involved with needy parent or children to deny them treatment unless we could come up with an answer as to how to handle the pressure coming from outside of the clinic and the pressure of the guilt feelings of the staff members. The former was easily dealt with by a directive from the director, but the latter did not so readily respond to directives. Another answer had to be found and that was to develop another kind of service which could be of some help to those in great need of help, though it may be far from sufficient to satisfy our own standards and our own drives to rescue everyone that comes to us for help. Furthermore, the staff had to be developed in their diagnostic thinking so as to gain information regarding treatability during the diagnostic process.

In the diagnostic process or, as the caseworkers choose to call it, intake process, it has always been assumed that each case will be a treatment case unless proven otherwise. This was changed to the opposite thinking; that each case will be seen only for diagnostic and consultation service, and only when treatment time is available for a given type of case will treatment be considered.

It became all the more important to pay attention to the diagnostic function of the clinic and to define more precisely what is to be accomplished in the initial intake and diagnostic phase of contacts with families. We have set up the following requirements of accomplishments during this phase.

1. To obtain a reasonably clear picture of the manner in which the family operates. If that picture could not be determined, the obstacles to that goal must be visualized.

2. A general picture of the psychodynamics of each member of the family.

3. An assessment of the strengths and weaknesses of the patient an those upon whom we are dependent for reality support in treatment, thus indicating what strengths we can ally with in treatment.

4. Indications as to what motivates the patient to come in for treatment, both consciously and unconsciously, and whether these can be used or changed so that they can be utilized to help the patient.

5. Indications that the demands upon the patients in the treatment process are explained to the patient and that he really accepts them or can be helped to accept them.

6. The ability of the patient to get involved into a treatment relationship such as is demanded by the type of treatment to be used.

The initial interviewing is assigned only to the most experienced staff members, whose diagnostic acumen is more mature. In contrast to earlier practices of being careful to not get patients involved, we deliberately attempt to involve the patient in a treatment relationship from the very onset in the hope that whatever information is given is given in a meaningful, feeling-full manner to which the interviewer can respond in a like manner with indications of acceptance, understanding, and possibly clarification.

This procedure serves several purposes: (1) It tests the patients' ability to relate and accept help as we need it. (2) It indicates to the patient that here is someone who not only demands information but who can also give understanding and through it help. (3) It allows quicker accessibility to the inner feelings which hasten the diagnostic process. It allows quicker accessibility to the inner feelings which hasten the diagnostic process. IT also has its disadvantages in that the staff members feel more obligated to continue or promise treatment and experience greater guilt in separating from the patient when they successfully involve the patient in a therapeutic relationship. Unless very skillfully handled, the patient also may be traumatized by either being refused treatment or upon reassignment for treatment to someone other than the original interviewer.

During the interviewing process the interviewer is constantly alert to determine the primary core of the problem both as to person and as within the person. For instance, who in this family most urgently needs treatment that would be of greatest benefit to the whole family? This person is not necessarily the sickest one, for the latter may be untreatable or too difficult to treat—it ma, in fact, be the health stone, for her presents the potentiality for pulling himself up and carrying the rest of the family with him. In a given individual what changes need to be made to obtain desired results? It the ideal result cannot be achieved which factor that is most amenable to brief treatment would most benefit him or preferably might so help to redirect the persons' thinking and feeling that further maturation can be expected?

It is expected that in many cases these questions can be answered fairly early, and indications of these answers are given to the patient also as a test of his ability to accept them., develop insight, and utilize them. This also serves to given the patient something that he often can very constructively utilize in his or the family's interest if further treatment is not available. The results in some of these cases that have had only from one to ten interviews is often quite astounding and demonstrate how they were able to carry on once their eyes were opened to some understanding of even only one facet of their struggle.

Through this change in he intake and diagnostic procedure, the latter has become a like, meaningful, highly responsible task instead of the routine presentation of information for an intake conference, and we do obtain much more useful information for the decision of selection of cases, and the patients are much better prepared for treatment once assignments are made.

In considering the selection of cases for treatment, we must first consider the basic function of the clinic. We basically exist to treat, to train, and to carryon research. Fortunately, the latter two functions serve materially to carryon the primary aim of treatment, but influence the selection of cases in that direction. It is inevitable that for each trainee a balanced case load of required variety be selected, and that certain types of cases be avoided at given times for a specific trainee or that he be assigned certain kinds of cases that would be specifically helpful in working out certain problems in the individual's training. We have tried repeatedly to define the kind of cases we would like trainees in the different disciplines to start out with, but have never been successful in meeting our criteria. Somehow the cases we considered ideal to begin training with are rarely available at the right time for the right person. We do, however, make every effort really to get acquainted with the new trainee before assignment are made so as not to overwhelm him at the outset by the assignment of cases which might activate serious conflicts within the trainee. Too often, we learn of the trainees difficulties only after the cases are assigned, with unfavourable reactions for all concerned.

Each research project needs specific case material, and selection therefore is simply determined by whether the cases meet the needs of the specific project being carried out. It is vital, of course, that it be determined beforehand that the staff participating in each project be competent to handle such case in treatment.

The selection of cases for treatment is furthermore complicated by the varied degrees of training and varied personalities of the staff of the clinic. It might be ideal to have a clinic with a fully developed, mature, psychoanalyzed staff that could take on any case, but in most clinics such as ours, most of the staff members are in various phases of training and experience, each one with their own limiting personalities which makes it essential that we not only know our cases well, but also that we thoroughly know the strength and weaknesses of each staff member prior to the selection of cases for treatment. In addition, we are also dealing with three different disciplines of varied training which again poses the problem of what skills may do best in a given case, or frequently the question must be raised whether the given staff person with the given skill and limitations will do at all. These are practical realistic factors that must be dealt with before each cases can be considered on its own merits.

In a public clinic, what criteria should one use in the selection for treatment? Or course, we have to determine that the case is treatable, or can be made so, but we have to turn way every many treatable cases. Shall we accept only cases, as he director of one clinic expressed himself, who have parent with healthy egos? We see some such cases in private practice, but very few come into the clinic. What a wonderful way of cutting our budget to a fraction of its present size. Shall we given preference to behaviour conduct disorders where there is no organized neurosis yet? Theoretically this can be justified, but in reality some of these, because they are tied to neurotic family members, may be less amenable to treatment then some neurotic family members, may be less amenable to treatment than some of the organised neuroses.

We select by certain symptoms either because of their importance or because they are most amenable to treatment.

But where is the case with just one symptom? A mother, for instance, may bring a child in because of enurisis. This is the presenting symptom so far as the mother and child are concerned, but is that all that concerns us? We must then dilineate what kind of enuresis and what goes with it, so that we may have to set up impossible algebraic or geometric formulae as guides.

So far as symptoms being amenable to treatment are concerned, we have all had experiences of unsuccessful cases, for instance, where the presenting symptom were given up after the first visit, and quite successful cases where the symptom may till persist by the time successful treatment is terminated.

If we try to set up criteria for selection on a genetic level, we are again in trouble, because as important as it is for us to know the level of functioning of our patient and level of genesis of his symptoms, this does not provide a clear line of cleavage which can serve as the sole basis for selection. The genetic considerations are vital in determining the method of treatment to be employed, though taken by themselves they are not nearly as much so in children as in adults; particularly is this so in the younger children.

Various devices have been employed by different clinics to limit selection of cases. This is relatively easy in cities where there are many clinics or the special purpose clinics that have no obligations to the community beyond what they wish to make it. For instance, the captive type of clinic within a school or hospitals of necessity limited in its cases, but the problem of selection for treatment nevertheless arises because of the time consuming aspect of our work. Age limitations are in common use, but within each age there are still a great variety of cases.

One clinic in this region has set up a system of grading cases from I to III on he basis of estimated treatment time needed, and balancing the case load with preference being given to the group with the least length of time required for treatment. We tried this system and have met with complete lack of success, in that we still tend to remain overloaded with long-term cases of dubious progrnosis; particularly, since this system requires turning away the group III cases (the group with the longest treatment time and the poorest prognosis) which the staff finds rather difficult to do.

In our long struggle with this problem of selection, we gradually evolved a scheme which I hope will work ultimately an thus far, as it has been applied, gives promise of doing so. In this scheme we have set down six simple criteria for the selection of cases for treatment, regardless of the diagnosis, symptoms, or ho good or bad the prognosis. The criteria to be met are:

1. The patient must be accessible to treatment.
2. There must be sufficient inner resources that can be utilized in treatment.
3. There must be sufficient environmental resources to support treatment.
4. A reasonable goal can be attainment in treatment in a reasonable, foreseeable time.
5. The goal chosen must be sufficiently important to the individual in our judgment and of sufficient help to be worth while for the patient.
6. There is good likelihood that the patient can maintain the gains to be aimed for after treatment is terminated.

This plan allows for a great deal of flexibility in case selection but also demands a realistic approach. It does not allow too rigid exclusion, nor the catering to impossible aims. It eliminates neither long-term nor short-term treatment. Its only demands that whatever is done be worth while to the individual; and if it is not so, that neither the patient's nor the staff's time and hopes be wasted by trying to do the impossible or not getting enough return in the time and effort.

For instance, most clinics turn away th intellectually borderline and mentally deficient. Why? Just because we cannot raise their intellectual level to normal? Quite often with the expenditure of a relatively small amount of time with the child and family, either in a short period of treatment, or in guidance over a prolonged period, or both, treatment may make a tremendous difference to that child, his family, as well as the community. Because we cannot make them top-notch citizens, should we not at least accept a lesser goal if it is attainable within a reasonable expenditure of time and effort? The same question must be raised in regard to some of the physically handicapped children. If a psychotic child can be helped to make some adjustment and hold on to a marginal function, should we not try? Under this scheme the answer would be yes only if we have the requirements for treatment, and that these goals can be achieved in sufficient time really to be of lasting value. The answer would be definitely no if it appears that at the end of a long treatment period the patient might be changed but no better off. We can decide to take on one case for long-term treatment because the goal thereby attainable is worthwhile, but in another case not. For instance, I recall a little twelve-year-old girl with a disturbance in which she would rock back and forth. This was embarrassing to her and the family. While this was a relatively mild symptom, I felt its roots were such as to require years of treatment and I did not feel this to be a worth-while undertaking. She did invest a year or more in treatment at another clinic without any gains. In contrast to this, I recall accepting for treatment a psychotic girl, though thee were great odds against treatment and though I anticipated that she would probably be withdrawn from treatment. I felt that there was a possibility even in a short time to reach her an help her to get a grip on reality which could possibly sustain her. We have a tendency in clinics to refuse to treat children when the parents are uncooperative and when we cannot hope to gain their full cooperation. Need we be so exclusive? Sometime, but not always. There are situations where despite all odds, sufficient help can be given to do a sufficiently worth-while amount of good. We do have to be sure that we are offering the patient a reasonably seaworthy raft, that he can hold on to, that will not be taken away from him, and that with it he can reach shore. We must not waste resources building for him a yacht or battleship he cannot get into, cannot use. We might be able to teach him to swim we might even have to carry him ashore, or we might just need to show the way to the shore. However, if he cannot swim, or we have o raft, he can not take the direction or could not last long if he should reach shore - why then make a demonstration of heroism when there are others who can be rescued in the meantime? Why appease our conscience by offering him a useless straw to hold on to?

Of course, it is very pertinent to this scheme of criteria for selection for treatment tat we be realistic in setting attainable goals and not chase after rainbows. It means that often, particularly in the long-term cases of dubious prognosis and the otherwise handicapped, we must be satisfied when a reasonable goal is reached.

Nursery School as a Child Guidance Clinic

The Nursery school referred to in this chapter is part of the Department of Psychiatry of the Western Reserve School of Medicine. The School has a capacity for fifteen children of preschool age who are suffering from various kinds of emotional disorders. Depending on the diagnosis, the children either receive treatment themselves (psychoanalysis or psychotherapy) or are helped via the mother, with whom the child therapist does a certain type of guidance work. This work ha been described by Mrs. Erna Furman in her article "Treatment of Under-Fives by Way of Parents"

This viewpoint makes present-day diagnosis consist of a meta psychological evaluation of the total personality. It seems to me that this new formulation is of the greatest importance for the evaluation of the conflicts of the child under five. The child under five gives us the opportunity to study two types of conflict, namely, the one existing between the infantile ego and the environment, and the other type in which the conflict has already become internalized. This diagnostic distinction between two groups, in which either the external or the internal conflict prevails, leads, as will presently discuss, to the decision whether guidance of the mother can be expected to be of sufficient help or whether direct treatment of the child will be necessary. Where the conflict is still an external one, that is, between the ego and objects in the environment, there of course we have to include in our evaluation all the parties concerned. It will not be sufficient to study the conflict only from the side of the child, but it will also be necessary to take into our consideration the attitudes of the objects, attitudes which are of course based upon the personality structure. Thus we have focus our attention upon the interplay between the child and his environment. With this group of children, a change brought about in the object may be sufficient to produce a favourable change in the development of the child.

It is evident that in those cases in which the conflict is already internalized, a change in the object will no longer be sufficient to solve the internal conflict.

Since, upon the admission of children, the mothers stay with them in the nursery for as long a period as we consider necessary, our nursery school setup in Cleveland offers an excellent opportunity to observe and to evaluate the interplay between mother and child.

For instance, one of the most frequent problems for which mothers seek help is separation anxiety. The origin of this problem can lie in traumatic events, in separations, such as the hospitalization of either mother or child, or the child might have a problem concerning aggression and therefore have the need to be with mother constantly, to reassure himself that mother I all right. On the other hand, it may be the mother who is anxious about her child and cannot let the child separate from her. Often the mother is not conscious of this, nor can she tell us about it, for this anxiety can be rooted in deep-seated conflicts within herself. Such a separation anxiety of the mother will of course induce insecurity and fears in the small child, and this situation often cannot be cleared up in observations of short duration. For instance, a single diagnostic interview that we would have with mother and child would not suffice.

Mothers bring children to us, seeking help consciously for a certain problem that may not seem of too great importance. It then often happens that our diagnostic study and observation in the nursery school reveal more serious problems which the mother has to deny, and she therefore displaces her worry to some minor difficulties. IF a mother uses denial as a defense against recognition of certain difficulties in the child, we can be sure that the child has taken over this same denial, and we have to rely on the objective observation in the nursery to recognize the real trouble.

Case Studies

Catherine was three years old when she was admitted to the nursery school because she seemed to her mother to have an acute reaction to the birth of a bay sister, now four months old. In the mother's interview with Mrs. Benkendorf, the mother said that her relationship with Catherine had become very saying about routine matters such as getting dressed, etc. Catherine had never been a very good sleeper, but she now refused to go to bed, would wake up during the night screaming, could not be alone any time during the day or night, could not bear to be separated from mother. At this point in the interview Mrs. Benkendorf asked the mother under what circumstances she had left to go to the hospital to have the baby, and it turned out she had left in the middle of then night without saying good-by to Catherine. The child had also gone back to wetting and soiling, although bowel and bladder control had been easily achieved at the age of two and had seemed well established. Catherine also needed her bottle very much again, although she used a cup when she went visiting.

The mother was very perceive of her child's unhappiness, tried very hard to be as giving as possible and reassuring of her love, but was puzzled by the child's steadfast refusal to grow up and to talk. This interview with the mother seemed to show beyond any doubt that the child's separation anxiety was due to the trauma of mother's sudden hospitalization, and her regressive behaviour was due to jealousy because of mother's attentiveness to the bay sister.

Let us now see the report of the teacher after Catherine had attend the nursery school six days: The child's adjustment to school has been a slow, arduous process, and although she has seen in school six days now, she cannot even stay through records yet.

It has been very difficult for the mother to allow Catherine to become interested in being part of the school or in doing things for herself. Mother hovered about Catherine for several days, pouring juice for her, telling her what to say and then saying it for her, helping her with her things in the bathroom, telling her when to go to the bathroom, etc. She could not keep away from Catherine, although he teacher continually requested that she stay out in the office.

A report on part of Tuesday morning will show how things went:

She began her morning in the waste room today. The first thing she had to do was to go to the toilet. She said she wanted to go alone. She crept into the bathroom and closed the door. After a while the door opened and she asked for her mommy. Her mother rushed in to help her immediately. When Catherine cam out, she thought she might like to blow bubbles, and she smiled just slightly once when I showed her how to blow to blow one.

After the bubbles, and she smiled just slightly once when I showed her how to blow one.

After the bubbles, which she just barely touched, she worked at the ironing board a long time (about ten minutes) with obvious enjoyment and contentment. When she finished, she ran to her mother for a hug, and her mother asked, "you'd like a drink of water, wouldn't you?" Catherine nodded. Mother said. "Then go get a drink. As soon as Catherine reached the fountain, mother asked, "Catherine, do you have to go to the toilet?" Catherine nodded again. Mother said, "Well, hurry, hurry," Catherine smiled, and it looked s though she were wetting her pants right then and there. She looked at her mother, and her mother said, "You want me to help you" (statement). They went to the bathroom together, and mother emerged with the wet pants. She did not seem at all surprised that the child had wet, although Catherine had been to the toilet fifteen minutes before.

At juice time it was suggested that mother sit in the office and that Catherine come and go as she pleased. Catherine ate and drank nothing and refused to answer when the teacher asked whether she was through.

At music, Catherine said she could come upstairs without the teacher. She poked her head in, and when I said, "Come in, join us," she became frightened or angrily resistant and started to sob. We took her down stirs where she became involved with her mother again. Mother could not stay in the office but found excuses to follow Catherine outdoors. Mother improved at lunch and stayed in the office throughout lunch.

The teacher became very direct with the mother and told her that the separation problem was hers and not Catherine's, and that it would be necessary for her to remaining the office if she expected Catherine to learn to stay at nursery school. Since then the mother has been doing a little better, but it is still very difficult for her to allow the teacher to do things for Catherine when she needs help.

Usually Catherine speaks only a word if she wants to make a point very badly, and this word seems to come out of her with terrific reluctance, Sometimes she will make only a grunting-moan kind of sound and expect you to understand her. On a few rare occasions she hands carried on what might be called a conversation—several words and phrases linked together and spoken with some animation and a bit less hastily. One subject with which she was able to do this was her daddy. She told me that she and mommy were going shopping for a birthday present for her daddy and that he would have a cake. This, of course, was not all told in one sentence but sort of coaxed out of her. However, her eyes were bright and her voice had some animation and lightness in it—something we do not often hear. Also, she laughs occasionally and smiles, which changes her whole personality. This seems to be about the only part left of her that is really Catherine and has not been superimposed.

Catherine wet herself purposely whenever she feels the urge. It can be fifteen minutes after having gone to the toilet, and it can take place while she is in the bathroom washing her hands or playing at the bowl. She stands still, a smile comes across her face, and then you know she is wet. For the first few days her mother had to help her almost completely in the bathroom, because she wanted to and because Catherine wanted her to want to. After mother began to remain in the office more consistently, Catherine began to expect as much from the teachers as she was used to from the mother. I have Catherine a little chair to sit on in the bathroom, and in two minutes the wet pants and overalls were off and the clean ones on. It was very apparent at that time that Catherine is willing to help herself and would quickly learn to do so if her mother could allow it. However mother also stresses the toilet a great deal, mentions and reminds her very often to make a pee-pee, and is generally very involved with toileting the child and changing her. Sometimes when Catherine has wet herself and we want her to change herself, she sys, "see, Mommy," and we know she is angry with us because, after all, her mommy will do it all for her, whereas we expect something of her.

The first time Catherine stayed for lunch, she at first refused to come to the table but sat in the office with mommy. The teacher asked her to pass some biscuits to her mommy and then to put one on her own plate. Catherine finally came to the table, and while everyone else was finishing desert or listening to records in the other room. Catherine ate helping after helping of peas, and nothing but peas. The next day Catherine ate a fairly good lunch of wieners, toast, and tomato, with no fuss at all. With the third lunch, however, e began pattern that she has followed ever since; she took a bite of egg-salad sandwich and held this bite in her mouth for at least fifteen minutes, although the teacher explained that she did not have to eat anything she did not want and could spit it out at any time. Catherine seemed to expect the teacher to reach into her mouth and take it out for her, but when she understood that this would not happen, she spirit out herself.

The mother had been watching this lunch process from the boot, and when she emerged from the booth her face was very smug and satisfied-looking as if to say, "See, they can't handle may child here as well as I do!" It is evident that mother is glad when the teachers fail with her child in situations where she succeeds, and she definitely has a need to prove that she can handle her child the best. Whether it is getting Catherine to go to the toilet (a feat on which she puts a high premium) or to eat her lunch or whatever, this feeling prevails.

Although we have felt that Catherine's adjustment to school has been very rough-going, we may be accomplishing a little bit, as evidenced by Catherine's behaviour toward her mother on one occasion. After lunch one day she was getting ready to go home and began to comb her hair. Mother took the comb from her and said, "I'll do it for you," Catherine aid, "why?" The when they were ready to go, mother said, "Okay, Bugs Bunny, let's go," Catherine said very firmly, "I Catherine," and turned to me and said, "I Catherine"

After this report, our assessment of Catherine' separation difficulties will be a different one. It was obvious that this mother had difficult in separating from the child or in letting the child become active and independent. And it was finally the child who had to tell and shown mother that she wanted to do things for herself and not be kept little.

12

LANGUAGE DEVELOPMENT: BEGINNING OF TWO-WORD PHRASES

Normal processes of language development; and in this chapter we shall briefly discuss the nature, functions and significance of play, nature and social and moral development. Language and play contribute in an important way to children's mental and social development. How vital a role language plays for cognitive expressions was stated in the previous two chapters. In the classroom language is used to instruct the child in performing the cognitive tasks and to socialise the child into the norms of the school and the classroom. For his social development, the child has to be enculturated through language because language is culture-bound and because "how we grow up, the kind of person we become, the ways in which we act are largely determined by the culture in which we are reared." The socialisation process is of utmost importance to the development of language and thinking.

A child learns a language to communicate with those around him in his environment. These surroundings – which we might call the child's linguistic environment –and the unique experiences he has in them have a profound effect on his language development and the mode of cognitive expression. How we think and speak mirror the environment in which we are socialised. A poor linguistic environment can have dramatic consequences for the child later in life in terms of not only social development but academic achievement as well. An impaired cognitive development may turn an individual into a social misfit, bereft of any sense of moral values. Thus, a child's linguistic environment is most significant with respect to his rate of language development and, it follows, his thinking, or cognition. Children proverbially socialise each other. A good illustration of social growth in children occurs in their play. Play at first is apt to be between two or three children. This condition may continue up till age 5. Gradually more children may participate until at age 10 a regular game with rules may be enjoyed. Play provides for the acquisition and exercise of social skills - co-operation, conversation, mutual respect, friendliness and courtesy. "Make believe" play, saturated with imaginative thinking, imaginative logic and egocentricity forms an inseparable part of the preoperational period, the first half at least, of cognitive development. Objects in the world about the child become active living entities with which he may identify himself. For example, the child is shown a picture of a lion and is told something about the animal's characteristics, or he sees a lion in the zoo.

There upon the child becomes a lion and stalks around the house, growling in true lion like fashion. Towards the end of the preoperational stage, "because the words are symbols, children use them to shape their world: as in Alice's Adventures in Wonderland, words mean only what the child wants them to mean, and the meanings may differ from those that you or we attribute to them".

Development of Language

The only thing that can be done is to discuss in some detail but with caution, the broad principles of language development in an environmental context. Most of the books on psychology and educational psychology deal with the acquisition of the English language on which a lot of research has been done. This is not suitable in the Indian context, thought references could be made to certain research findings.

Many tongues are spoken in India. The languages spoken in India behave differently from one anther and are fundamentally different in their characteristics from the English language which does not occupy the same position in India in the present that it used to about four decades ago. It has become the third language to be learned in some of the States of India and in some others, remains only as an optional subject in the school curriculum. There are a few English-Medium schools, no doubt, but their number is too less to be taken into consideration.

The bones of the English language are of three kinds:

Word-order. Word-order in English is fixed. If the word-order changes, the meaning also changes or turns meaningless. For example, "The lion killed the tiger." "The tiger killed the lion." Because word-order in English is fixed, the models for the different kinds of English sentences are fixed also. The languages in India do not have a fixed word-order and fixed sentence patterns.

Use of Structural Words. In English, words do not usually change when they are put into sentences. Instead, English uses small helping words like *do, will, to, at, from,* to show the change in meaning of the main words. These words are called *structural words,* that is, the words we talk with, because they are necessary to the structure of the English sentence. In the sentences "Did he go?" "Will he go?" "does he go?", the verb go does not change; but a different small word—*did, will, does*—is used to show the change in meaning. Similarly, in the phrases—at the house, to the *house,* in the house, from the house—the chief word *house* is not changed; but small words like *to, at, from* are used to show the change in meaning. In most of the languages of India this is not the case. In the sentences, the chief verbs or the chief noun words themselves get changed or inflected.

The Use of a Small Number of Inflexions. English also uses a few word changes to show various meanings. For example, "he *goes*" "I *answered.*" Or in nouns: One *boy*, two *boys*, and in adjectives and adverbs; *quick, quicker, quickest.* But English uses a very small number of *inflexions.* On the other hand, most of the languages in India are highly *inflective* and highly *phonetical.* For example, not only do the verbs get inflected for person, number, and tense, but gender and case also.

The chief noun-words themselves get inflected to show change in meaning. And, the characteristics of a language, the mode in which a language behaves, largely determine how a language develops in the child, or is learned by the child, an issue about which psychologists argue. However, as the bulk of research in how English as the first language is learned has already been done in most of the advanced countries of the world, one has to discuss it first before one goes to cross-cultural studies in language development. The only property common to all the languages in the world is the definition of language itself. What is language? "To the philosopher, language may be an instrument of thought; to the sociologist, a form of behaviour; to the psychologist, a cloudy window through which he glimpses the workings of the mind; to the logician, it may be a calculus; to the engineer, a series of physical events; to the statistician, a selection by choice and change; to the linguist, a system of arbitrary signs". But virtually, language is a complex system of systems. It is a system of sound signals, a system of written symbols, a grammatical system, and all these three systems, are hitched together to generate a system of meaning.

Pre-linguistic Stage "In one study (Eisenson, Auer, and Irwin, 1963), the vocal sounds made by the infants were divided into five substances: undifferentiated crying, differentiated crying, babbling, lallation, and echolalia" (DeCecco & Crawford, 1977, p. 114). Babies' cries are very effective at gaining attention and their smiles too. Thus, right from the beginning the infant is equipped to give messages and to receive messages also. In the first month of life the infant cries in response to any non-differentiated source of discomfort. In the second month the infant cries in order to get persons to attend to his needs of food, movement, warmth, sleep and so on. During the third month he begins the cooing that gives way to bobbling in the fourth and fifth months.

Bruner differentiates three major ways in which caretakers of babies are able to interpret babies' intentions. The first is from the nature of the babies' utterances, which Rick's (1975) research proves. Secondly, from the accompanying movements, that is, the pattern of the babies' movements, which the research of Richards (1974) proves. And thirdly, from the actual situation or the context in which the interaction between the babies and the caretaker is taking place. For example, the caretaker will take into account how long ago the infant was fed, his position and so on.

The first speech-like sounds babies make are babbling sounds, which start whole the baby is about three months old and is in its full swing while he is about five to six months old. Summarising the research findings of Gregoire (1937), Piaget (1932). Stern (1928), Leopold. Kaper (1959), and Gesell (1940), Mackey (1965, p. 105–106) has recorded that while the baby is three months old, his level of understanding constitutes: Recognises human voice. No memory. Bound to present. Little co-ordination between grasping and mouthing; and the nature of utterance: Social smile. Spontaneous, disorganised sounds. Babbling. In the opinion of Siann and Ugwuegbu: "First, babbling gives the child an opportunity to gain control of the speech organs and, secondly, *babbling is very similar in all speech communities all over the world.*"

About the sixth month lallation begins—the child appears to be reproducing his own speech sounds. The child recognises friendly and unfriendly tones and produces many vowel sounds. About the seventh or eighth month the child engages in echolalia in which he imitates the sounds produced by others. When the infant is about nine months old, he reacts to gesture and shows preference for certain words. All varieties of sounds and sound combinations appear at the utterance level.

One-word utterances. Somewhere round the child's first birthday his first words appear. First, these are one-letter words, in some of the Indian Vernaculars, and subsequently the child utters two-letter words. Thus, first he utters sounds like *ma ba, pa, da, oo, ee, aa,* and than goes to *mama, baba, dudu, papa, aaoo, ooaa, doo daa,* and so on. When the parents hear some sound combinations they take them for words, and by acting as if they had meaning for the infant, they help establish associations between these sound combinations and certain persons or objects or actions. Thus, when the parents hear the child repeat syllables like *ma* and *ba,* again and again, interpreting the former as *mama,* the father will repeat it, while taking every opportunity of pointing to the mother as he does so and, interpreting the latter as *baba,* the mother will repeat it, while taking every opportunity of pointing to the father as she does so. Sometimes these words are used very broadly. For the baby, *mama* can mean any female person and *baba* can mean any male person.

But eventually the child, by now used to imitating himself, will initiate others, repeating their words, while pointing to persons or things. This pointing-and-naming stage soon develops into the naming-and-getting stage. These first single words are used for a number of functions: for description, for command, and for reporting. An example of the last of these three functions is that a child may see a bird and say "*koo*", track it across the sky by pointing and when it disappears, say "*tata*", So it seems that for the young child these first single words are not just isolated words and labels but are instead an attempt to express more complex ideas. When he is about 12 months to 17 months old, the child responds to simple commands at the level of understanding and at the level of uttrance, there is verbal play. It is the jargon stage. And the first few words are conditioned responses. "The child always learns his meanings through contexts of situations. The more the context comprises, the greater the probability of his guessing the right meaning. Although sounds are learned by imitation, meanings are acquired by association and developed through reinforcement".

Stage one speech: Beginning of two-word phrases

When the child is about 18 months old, his level of understanding is:

(a) Points to simple things on command, and

(b) Follows simple direction. His level of utterance is constituted of:

(i) Can speak about a dozen words or more.

(ii) Verbal behaviour is accompanied by geture.

(iii) There is beginning of language as combination.

(iv) Jagon begins to disappear.

(v) There is beginning of two-word phrases.

(vi) Vocabulary begins sharp rise (Mackey; Ibid; p. 106).

The infant begins to differentiate in the situation between elements like actor and action, and to combine these into two-word phrases like the noun-verb combination.

Those children whose first language in English, talk in telegraphic speech. They omit the structural words like articles and prepositions and use noun—verb phrases, such as "see *mama*" or "doll come". In the child's early speech he is intent on getting his meaning across and this accounts for the similarity to telegraphic speech. Thus the *first characteristic* of stage one speech acquisition in English is that children's language is much simpler than adult language and concentrates on expressing meaning.

According to Dale (1976), we can understand the sematic relationships of stage one speech reasonably well if we analyse it in terms of agents (something that act), objects (something that agents affect) and *action*. So, if a child says in English "doll come", then doll is the agent. If he says "hit ball', then ball is the object. The child can express other sorts of meanings like possession with two words, as for example, "Dada shoe", "Mama ring". The *second characteristic*, then, is that it is often helpful to analyse stage one speech terms of *agent, action, object*.

As stage one speech develops, the number of words in the noun-verb phrase increases. He moves from the two-word stage of "hit ball" to "I hit ball" and then later to four-word sentences like "I put dollie box." Thus *the third characteristic* is utterances get progressively longer.

At this stage, however, the child has not yet attained the ability to use *inflections*, and the use of articles and prepositions. So he will say, "I give Kitty milk" not "I am giving Kitty milk." Therefore, the *fourth characteristic* is that children do not use inflections such as are used in mature adult speech.

The first and by far the most important rule of the English language is that it has a fixed word-order; and from the beginning only, the English children learn it. They always use the agent, action, object form –"I hit ball", not "hit I ball" or "I ball hit." The *fifth characteristic*, then, is that from the beginning only children tend to use a fixed word-order.

"How well do these five points apply to the acquisition

of other first languages? Stobin has collected accounts of early language development in forty languages and he concludes (1979) that stage one language is very much the same all over the world' (Siann & Ugwuegbu, *Ibid*; p. 31). But except for the fifth point, of course. Does word-order tend to be fixed in the acquisition of the first languages of India? The answer to this last point is "No." In other aspects, however, as Stobin (1979) does, early utterances in the various languages studied read like direct translations of one another.

State two speech: Learning the use of grammatical points

When the child is about 2 years old, he understands commands referring to present situations and he may understand several hundred words. At this stage, jargon is dropped. Languages as communication begins. Great individual variations in pronunciation is observed. Uses vocabulary in the hundreds and uses pronouns. The child verbalises wants.

By the time the child is three years old, he understands questions referring to past. There is a large increase in vocabulary and gets structures right. He uses plurals, tenses, parts of speech and sentence patterns. He narrates past experiences and uses language as a substitute for action. (*Ibid*).

When he reaches his fourth year, the child possesses a great range of understanding but still confuses periods of time. His speech skills are perfected and there is no baby talk. On the other hand there is rambling torrent of speech. This is the age of *loquacity* (*Ibid*).

At five years, the child understands past events of childhood. He may understand over two thousand words in his mother tongue. He gives briefer and more critical responses. He begins counting and printing. He uses more complex sentences and has command over large vocabulary (*Ibid*).

In stage two speech, children whose mother tongue is English, begin to use inflections. Two important points characterise the development of inflections and sentence patterns on account of the irregularities found in the English language, in inflecting words; and because word order in English is fixed, and because word-order is the most important thing in every English sentence, the models for the different kinds of English sentences are therefore, fixed also.

(a) All children learning English tend to learn the inflections and sentence patterns in much the same order.

(b) They follow the same processes in learning inflections and sentence patterns and tend to make the same sort of mistakes. The second point leads us to the phenomenon of "*over-regularisation.*" That is, the child learns a rule and then over regularises it.

It is interesting to look at the process of over-regularisation in inflecting words. There are irregular verbs like *come* and *break* in English which do not take "*ed*" to form the past tense, as most English verbs do; for example, walk/ walked, talk/ talked. The child will first learn the past form of "come" and "beak" quite correctly, by imitating his parents, and will say "came" and "broke." Then he will realise there is a rule for the formation of past tense and will revert back to saying "*comed*" and "*breaked.*" Similarly, by analogy, he may say, "one boy"—"two boys" and "one man"—"two mass." One he has "over regularised" in this way, the errors are resistant to change. This tendency to pay attention to rules indicates the child is searching for patterns as he acquires his language. In the words of Mackey, "By now the child is continually making use of analogy, continually trying to apply a rule, as it were, always trying to find the regularly recurring elements of the language. And here again his analogy may not always be right, for he will regularise all plurals to produce uttrances like *tooths* and *foots* and invent past participles like *bringed* and *teached*. Here he enters the realm of grammar, where he must learn to make further distinctions."

All children throughout the world, as different research findings show, seem to learn the inflections in much the same order, and the phenomenon of over-regularisation is found. This may be, perhaps, on account of as Mackey (*Ibid.*) puts it: "The child learns one meaning at a time. The first meaning may not always be the most usual. It may even be a rare, figurative or idiosyncratic meaning. But it is the meanings which best fit his needs that survive. Having often more meanings than words, however, more things to say than means of saying them, the child extends by analogy those meanings of words which he already knows."

Stages two speech has two fundamental aspects. First, *function*—the child is continually striving to "express his own ideas, emotions and actions". Secondly, "the *structure* as well as the function of language become important". This means that the child starts to learn the structure of the mother tongue in a rule governed, systematic way, event-valley for the mastery of the grammar of his own language. Adults are sensitive to children's need to communicate and help them to acquire language by using suitable speech patterns. The older the children grow, the more the structure of ‘heir speech becomes like adult structure. In this learning of structure they show evidence of a need to search for patterns. It is through analogy that the child learns the grammar of his language. But he does not usually acquire a mastery of all its forms until after he has entered school. *How Adults Talk to Children.* As adults, when we talk to children, specially young children, we tend to talk to them rather differently. This is because we know that their understanding of language is more limited than our own; and yet, we want they should understand what we are saying to them.

"Adults seem to alter their speech when talking to young children in three ways—they slow down, they use short, simple sentences and they repeat themselves frequently....

Ferguson (1977) summarises a number of studies from all over the world—ranging from Morocco to Greece—that indicate that adults everywhere do take into account the fact that young children have less language experience and so need a more specialised kind of communication than other listeners do" (Siann & Ugwuegbu, *Ibid*; p. 33). In India too, we adults take the same approach.

Later growth in children's language: Refinement of distinctions in grammar and meaning. During two years of their age, from 6 to 8, children completely understand everyday utterances. They begin reading, thus increasing passive vocabulary. They have complete speech habits of the language. The only need is reinforcing them through practice. The children hold on to patterns learned.

The joint aspects of function and structure in language become more and more elaborates as children get older. They are qualitatively able to speak. They can remember the sequences that gives them a basis for their mastery of sentence structure. Plurals, word endings and different parts of speech begin to appear. Simple sentences become more and more complex. They have most of the basic elements of language and their language becomes more similar in structure to adult language. However, the child has to increase the number of the basic elements of the language at his command and refine his distinctions in grammar and meaning. We may have to wait until after the age of ten to twelve before we can expect the child to understand grammar as grammar. It is generally after the age of ten that the unilingual child begins to make a distinction between a word and its context; yet, at this stage, his use of language extends from being mainly used in communication to becoming a very effective tool in his growing understanding of the world around him. "We can now sum up by enumerating the main procedures involved in learning the first language. They are: (1) trial-and-error, (2) rote memory,(3) imitation, (4) association, and (%) analogy".

Nature, Function and Significance

What is Play? From a baby in the cradle up to a doddering old man one foot in the grave—everyone does play, with endless varieties, of course. Play activities may range from random activities to highly organised games, from shaking a rattle through cricket to playing a game of chess. Education is a human enterprise, but not play. Not only do humans play but also primates like monkeys and chimpanzees, as well as other mammals like cats and dogs, you must have observed two little kitten or puppies tumbling over each other in the sunshine. Play is universal. Young ones play the most—a Kitten, a puppy, a human child; and the human child has the longest period of play of all.

Children would not be children if they did not play. Childhood is the play time of life. Through play the child gives a free, natural and spontaneous expression to his nature and needs. It is in the many varied forms of play, may get revealed the possibilities of his future development—emotional, cognitive, social and moral. A great philosopher-educationist seems to have said: "Childhood is play personified". Yet, one wonders why we try to philosophise play more than we try to psychologise it. Very few modern authors of textbooks on psychology and educational psychology have given a place to "*play as play*" in their work whereas in a number of books on philosophy and education one finds play to have been given a good treat. And an excellent example of this is what Phenix in his famous book—"*Philosophy and Education*", 1966, p. 249—writes about play.

"The typical instance of play activity is participation in games. The essence of a game is the set of rules which define the choice open to players and the rewards or penalties resulting from various developing circumstances. Implicit in this are three characteristics of play which are worthy of emphasis. First, play is associated with human freedom...Play can never be compulsory. It is essentially a manifestation and an exercise of personal freedom. Second, his freedom operates within the context of agreed upon principles...Thus law and order are of the essence in play. Third, this freely chosen pattern is in effect a new miniature world created by human artifice. The world of play is one of makebelieve...

Considered in this manner, play is clearly of the highest significance educationally."

But do organised games only constitute play? The answer is "No." "One student (Parten, 1932) described well the stages in play as follows:

(i) unoccupied behaviour,

(ii) onlooker behaviour,

(iii) solitary independent play,

(iv) parallel activity,

(v) associated play, and

(vi) co-operative or organised play. Note that associate play comes fifth in the series and later in time".

Catherine Garvey in her book on *Play* (1977) has suggested that the following five attributes are commonly accepted as being characteristic of play.

(a) Play is enjoyment. It is something players want to do

(b) Play has no other goals than enjoyment. It is, *intrinsically* rewarding.

(c) Play is freely entered into. Players choose to play.

(d) Play is active.

(e) Play is related to other activities.

Phenix and Garvey suggest almost the same thing. But their suggestions do not answer the question—"Why do children play?" - adequately and uncontroversially. A kitten plays. A puppy plays. A chimpanzee youngster plays. A human child plays. Therefore, Siann and Ugwuegbu point out: "This suggests, it (play) Has a functional significance and indeed it is clear that in play the young of many species are able to experiment with different kinds of behaviour. Although adults do play as well, it seems that the capacity for play, as we have discussed it, declines with age. It seems appropriate then to look at the role of play in child development."

Why do children play? So, there is a role of play in child development. But undoubtedly, children are not conscious of this fact. Then why do they play? Long back, the engineer philosopher Herbert Spencer argued that play was the expression of a *surplus* of *energy*. The young creature, being fed and defend by its parents, has more energy that its needs for maintenance and growth, and it expends the surplus in play. This explanation of the phenomenon is hopelessly inadequate. It does not explain why play takes definite forms and why we play when we are tired. The same criticism holds good for the explanation offered by Maritz Lazarus who thinks that play is *relaxation* and brings *recovery from fatigue*.

Karl Gross believed that play was nature's way of preparing the individual for adult life, a period during which instincts must be adapted to the activities essential to the well-being of the adult. Hence children played at adult activities—keeping house, running a store, and tending a sick doll. For him play is *teleological* and play is *anticipatory*. Thus, to Gross, through the multifarious forms of play the child rehearses the complex and serious activities he shall engage himself in, in his adult life. Play becomes Nature's mode of education, in the sense of adaptation to environment. In his play the boy is unconsciously 'trying out' various possible occupations of manhood. Professor Stanley Hall passes a pungent and severe remark against the explanation of Groos when he says that it is "very partial, superficial, and perverse," since "it ignores the past where lie the keys to all play activities." But doubtless, there is much in his objection. According to him, the child is not so much rehearsing the serious activities of his own adult life as harking back to and recapitulating those of his remote ancestors". Thus in the hands of Professor Hall, play becomes *reminiscent rather than anticipatory*.

Professor Hall believed that just as the human organism recapitulates the biological evolution of the species, so each individual in his social life *recapitulates* the cultural evolution of the race, through play. There in his two-volume work on *Adolescence* (1904) at different places he writes, regarding play "as the motor habit and spirit of the past of the race, persisting to the present, as rudimentary functions sometimes of and always akin to rudimentary organs....Thus we rehearse the activities of our ancestors, back we know not how far." And again: "The heart of young goes out into play as into nothing else, as if in it man remembered a lost paradise." One example will suffice. Hide and seek is reminiscent of the time when man had to slip stealthily through the trees and tall grass for concealment. Hall's theory was further expanded by George T.W. Patrick. He worked out a complete parallel between play and the chief types of life, habits and occupations during the stone age.

McDougall sees in play the expression of the instinct of rivalry. In supporting McDougall, Ross explains: "Play is *cathartic* in its action; that is to say, it provides an outlet for certain pent-up instincts and emotions which whether in childhood or in adult life, cannot find sufficient direct expression. In civilised life the instinct of pugnacity, for example, finds insufficient scope for exercise. By nature we are fighters, and fight we must, so that civilized man fights in play. Every game is a sham fight, in which no blood is shed, or even anger displayed, but which nevertheless relieves the energy of this instinct, providing a vicarious channel for its expression".

While some authors are of the opinion that the *theories of play* discussed so far are complementary rather than contradictory, others are of the opinion that the statements of these writers can hardly be called "theories" and if at all they should be treated as "theories", no one theory fits all types of play nor can a single theory explain the fact of variation; for each of the writers selected only one characteristic of play and ignored others, and sought to describe all in the light of the one characteristic.

"More recently, largely within the last quarter of a century, sociologists have turned attention away from the purely individual emphasis to the value of play as a medium of social interaction. Play develops from the desire of the child to achieve ends which he cannot achieve alone, his desire for group association, and the consequent wider self expression...The artifacts of play are miniatures selected from among those of the adult world." Says Brown, and again he writes, "Sociologists are interested in play as a preparation for adult life, not, however, because of a possible carry-over of a specific skill. They see in play, and especially in its group aspects, the socialisation of the child, sensing the meaning of a "we-feeling," finding his role within the group, and learning the significance of adjustment, opposition, co-operation".

Making these statements, Brown seems to be trying to reconcile the doctrines of anticipation, recapitulation and catharsis. The play of children may, at one and the same time be the recapitulation of ancestral activities and the anticipation of the adult activities of the individual himself, provided that the latter are what the race has done for countless generations, and the release or discharge of the energy generated by needs in the direction of both real and substitute goals-energy that would otherwise be bottled up creating a tension reservoir in children, turning them maladjusted in life situations.

Now the time has come when we should examine the

functions of play in promoting development of the human child.

Play for mastery of physical activity. Children are very active begins. To be active is their natural and inborn ability and desire. Even through they are not aware of it and intentionally perform these activities, right from their early infancy their activity often serves to increase their motor skills. We have already noted in chapter eight of this book the close interplay between motor activity and perception in infants, throughout sensorimotor stage of cognitive development. Through countless experiments the child tries to discover relationships between sensations and motor behaviour, thereby eventually developing his motor skills. He grasps a spoon, looks at it, bangs it on the dish, and pushes both the spoon and the food dish to the edge of the table to know what happens if he does so. Before long he manages to fit different objects into every activity into which it is possible to put them.

This early tendency to play using motor skills increases with age. Infants trying to climb up and climb down steps on all fours, standing on the verandah and jumping down onto the floor below are an everyday affair at home. Older children running—one chasing the other—or tumbling over each other or wrestling, and other games involving motion are a very common scene both at home and school. You must have observed a popular play activity along the streets of most of the towns and villages of India. Young boys use an old bicycle tyre or a spokeless rim as the play-object. Using their hand or a small wooden stick, even an iron rod, they make the object roll and run quickly after their rolling toy. Very often obstructing road traffic. The fun of the game is to keep the rolling toy from twisting crazily when it his some uneven road surface. Hakes (1972), p. 28) observes this same play item very popular among the children of northern Nigeria.

Play, Exploration and Novelty. Everyone of us is fascinatied by novelty and more so the children. The children explore new situations so that they can progressively understand the world surrounding them. Corinne Hutt, as Bruner (1974) has recorded, demonstrated through an experiment using a 'super toy' designed by her for children of 3 to 5 years that the tendency to explore in early childhood is linked to later creativity.

Garvey tells us how exploratory activity develops into play, describing what happened when a 3 year old boy encountered a large wooden car in a play room for the first time.

However, children do not need to have toys in order to explore. The tribal children in India, as in African countries , roam the countryside and learn to distinguish a great variety of birds, animals, insects trees, grasses, fruits and flowers. Their interests bring them in contact with these things since the latter constitute the furnishings of the play activities of the former.

Whatever the nature of environment the childen come from be, it is essential that adults accept that the children need the opportunity to explore their environment and learn how to find things out for themselves. Helping the children gain new experiences, giving them the opportunity to manipulated things and explore their environment, we lay a good foundation for an intellectually alert adult. Thus, "the *heuristic method*, first applied to the teaching of science by professor Armstrong, seeks to put the pupil in the position of the discoverer, and to give him the elation of "some watcher of the skies when a new planet swims into his ken".

Play for Social Learning. Descriptions of preparation for adult life through play are very common in the cross-cultural literature. As in India, in African countries too, play often becomes an important part of informal education, especially in tribal societies, that characterise such societies. These descriptions reveal that the tribal children of both India and

Africa engage themselves in identical play activity not only reflecting the mode of adult life, but acting as a medium through which children acquire social, cultural, moral and religious values of their society. An excellent example is given below.

Castle in writing of East Africa notes: "Imitative play was an important part of information education. Boys played with wooden spears, bows, and arrows and shields made from banana bark; they built model huts and cattle pens, for these would be tasks for them when grown up. The girls made dolls, played at 'husband and wife.' Plaited baskets of grass, ground corn like their mothers, made little pots of clay and cooked imaginary meals." These items of play activity are found among children in most of the rural communities of India in spite of modernity and mechanicalism penetrating into most of the villages.

Fortes (1976, p. 478) writes about play activity found among children of Tale in West Africa. He describes how groups of girls and boys formed together to play at *house making* which one would witness in both rural and urban communities of India too. His description of how the playful activity of hunting among children shaded gradually into the real activity for which it was a preparation can also be witnessed among the children of most of the village communities in India. Plays is also the medium through which children acquire the values of their culture and are absorbed into its spiritual life. In India, rituals take multifarious forms both among the tribal people and the non-tribals. The philosophy of religion in our country ranges from fetishism through polytheism and henotheism to monotheism. The mode of worship differs from community to community which gets reflected in the play activity of children belonging to different communities.

Performance of various *pujahs* both at home and at school, in the tribal zones and non-tribal zones as well, stands a vivid illustration to this. Fortes describes about similar ceremonies performed by children of Taleland in West Africa. A value is a socially desirable objective. Play also reflects the dominant values of society. In India, marriage has a socio-cultural value. As Ralph Linton defines it, "A culture is the configuration of leaned behaviour, and the results of behaviour, whose component elements are shared and transmitted by the members of a particular society." The word configuration, as we have noted earlier in the second chapter of this book, suggests a sort of pattern. "The term *culture pattern* has come into common use in the years since Ruth Benedict wrote her well-known book called *Patterns of Culture in 1935*". "A culture pattern can be taken to mean the ways of behaviour associated with any permanent need or function in the social life" (*Ibid.*), and thus, the method of bringing up children in a given society, religious practices, methods of doing business and trade, ways of eating, drinking and performing marriage ceremonies, or ideals of life are examples of culture patterns which may differ from one society to another.

Almost at every home in the State of Orissa, the author has observed children engaged in a play activity which should be translated as "Dolls' Marriage." He thinks, in other states of India too, children play this item. The play is this: The children divide themselves into two groups, in each group both boys and girls, as the doll-bride's party and the doll-bridegroom's party. Beating drums on empty tins and playing on hollow-bamboo pipes the bride groom's party with the doll, bridgegroom sitting on an wooden box, thought to be the motor car and both dragged and pushed by one or two children, arrive at the doll-bridge's home. The bride's party treats the groom's party with respect and hospitality. A fake-quarrel often starts on the plea that the former party failed to observe the required customs and manners. Then the marriage function and the feast start simultaneously. Toy-earthenware are used as utensils and snacks and biscuits served to the members of both the parties. The chanting of mantrams is mimicked and the marriage function closes. Another fake quarrel starts over the nature and adequacy of 'gifts' the bride and the groom should get. The quarrel gets resolved, and together, both the parties with the bride and the groom start for the home of the groom, to have another feast. To illustrate how dominant social values get reflected in children's play Siann and Ugwuegbu (*Ibid*; pp. 35-36), refer to children's play in three different cultures. The first is the Tangu children of New Guinea playing a game called *taketak* and described by Burridge (1976), the second is a game called *ekak* played by the children of Nigeria and described by Uka and the third one is the children of Britain playing *Monopaly*.

Play, Cognition and Language. Manipulation and exploration of physical objects adds to the child's knowledge and mastery of the external world. In European countries, "*educational toys*" are designed to promote children's spatial ability—important in the study of physics and mathematics. However, children can gain their knowledge of shape and form just as effectively being encouraged to play with materials at hand such as pouring water back and forth from one can onto another stacking cans, playing the syphon system with two cans, water and a tube and so on.

Asking riddles, which is a very common game in India, helps children to think analytically. Riddles are called "*Paheli*" in Hindi and plenty of books in all the Indian vernaculars on riddles are available for the children to play language games and enjoy the fun. Children also use puns, called "*Doo Arthi*" in Hindi, to create humour or crack a joke among themselves. This is another language game closely associated with play. Practices similar to those as were prevalent in the tribal Zones of India by 1960, were also present in East Africa as reported by Castle; and as Siann and Ugwuegbu remark, these practices "are of undoubted value in the cognitive development of the child not only in the area of reasoning but in the area of language development as well." Castle reports:

"By the evening fireside tribal legends and proverbs were told and retold; there were riddles to test childish judgement and myths to explain the origins of the tribal and the genesis of man and the heavenly bodies. These, told with care and with much repetition, were the African child's education in what was often a complicated and beautiful language. There were no grammar books, no writing, but correctness of speech, so charaeristic of illiterate Africans, was learnt by imitation of their elders." This author firmly believes that what Castle reports of the tribals of East Africa even now holds good in so far as the more isolated tribals of the whole world are concerned, not only of East African or India.

Play and Creativity. In India, in all the tribal societies and village communities as well, music and dancing are very much a part of life and children joint in as the grow older because all children like to chant rhymes, dance and sing. Unity in diversity is our philosophy and ideal of life. Not only India is a multilingual country but also multicultural. One has to admit this fact. Thus, songs and dances, folkways and mores differ from community to community and state to state in India, the splendour and glory remaining the same.

Writing about the Gikuyu of Kenya, Keenyatta notes that children "merge insensibly into the dances of later years....". Sometimes chants and rhymes are woven into children's games, lana and Peter Opie, writing about children in Britain, describe street-games; many of these have rhymes and chants which are an indispensable part of the fun. In West Africa, Fortes reports that organised games involving dancing and singing are part of the traditional play of early childhood. "Children play them for the pleasure of collective singing, rhythemical physical activity and sensory and bodily stimulation". Encouraging children to express themselves in rhymes, dance and music can pave the way for creative work in this area. In most of the States of India older children play a language-game known as "*Padyanta*."

The game is competitive and an admixture of recreation and seriousness that promotes both memorisation and creative thinking. Boys or girls divide themselves into two groups, four to six members in each group. One of the members of the first group starts the game reciting a couplet or stanza of a poem, composed by someone else or by the member himself or herself. The second group—one of its members—must supply with a matching couplet or stanza, beginning with the last unit of sound the couplet or stanza recited by one of the members of the first group ends with. This is the rule and sequence that must be followed throughout and that group which fails to do so, loses the game.

One of the most interesting and important form of play is "*make-believe*" Garvey (1977) reports that there is evidence linking fantasy and make-believe play to a high-level of creativity. Gross, as we have seen, explains make-believe play as an unconscious preparation for adult life. Thus, make-believe games allow for role-playing and also for children to project themselves into the future; but perhaps a more satisfying and convincing explanation can be found on the lines of the Cathartic effect. This leads us to the belief that make-believe play and fantasy have another role, and a very important one, in child development and it is to this that we turn now. The Interaction of Play with Emotional Needs. The child in his usual surroundings does not finds sufficient scope for the satisfaction of all his emotional needs, notably his psychosocial needs of status, mastery and independence. His environment, physical and social, is rather too much for his limited powers and he cannot dominate it as he desires. In his make-believe play he finds an imaginary world wher he is important and powerful and where everyone does what he says, thus finding a channel for the relief of the pent-up energy of his emotional needs. As an imaginary bus-conductor or a train enegineer or a teacher he achieves the positive self-feeling for which his real environment gives him insufficient opportunity. In all make-believe play, then, we must look for a discharge of the energy generated by the emotional needs that would otherwise be bottled up. Most children, at one stage of their development, live in a world of fantasy peopled by creatures of their own making. Because of this we should not dismiss as lies the incredible stories that children may tell us about themselves. They may not be objectively true, but all the same they are not invented by children in order to deceive us. Make believe also helps the child to relieve exciting or emotionally important experiences, as after visiting a zoo where he sees a lion, he comes home and acting a lion, stalks around the house growling in true lion like fashion; or. After visiting a railway station, he comes home and is stimulated to become a "choo choo" train.

This aspect of play is sometimes used to cure children who are emotionally disturbed. This is called "*Play Therapy*" Play therapy stems out of "*psychoanalysis*" which, as stated in the second chapter of this book, refers to a method of interpreting and trating mental disorders, especially psychoneurosis. The underlying idea is that by representing a very disturbing experience or situation symbolically, and by returning to it and perhaps reversing what happened, the child will be better able to deal with the problem in real life. Toys, water paints, drawing materials, guns, blocks are made available to the child.

13

CONCEPTS AND PRINCIPLES OF PSYCHOMOTOR SKILLS

Technological advancements have changed the socially desirable objectives; and together, they have changed our culture. We ardently desire to enter into international competitions in games and sports, gymnastics and boxing, cinema and circus. Entering into these competitions is not possible unless we develop in our pupils from the beginning only psychomotor skills and promote their physical fitness by different methods and techniques. Without the development of perceptual-motor skills, children would never be able to learn how to walk and talk, run and jump, hop and skip, or dance and sing.

Even the basic activities of reading, writing and speaking would be impossible because these too depend on perceptual-motor skills, these are no doubt verbal skills and mainly perceptual, but all the same, they involve the motor component which is really important. Skills develop as integral parts of the individual's total development. A skill has three characteristics, say De Cecco and Crawford: "It represents a chain of motor responses; it involves the co-ordination of hand and eye movements; and it requires the organisation of chains into complex response patterns". There is a distinction between a skill and a psychomotor ability.

According to Fleishmen, (1962, 1964), a psychomotor ability is a general trait of the individual which is related to performance of a wide variety of skills. Operating a lathe machine, or assembling electrical components or flying an aircraft are specific skills. A skill is performance of a specific task. Manual dexterity and motor cc-ordination are basic psychomotor abilities. Whereas manual dexterity is needed in assembling electrical components, motor co-ordination is needed piloting an aircraft.

"Motor skills have form, internal organisation, rhythm, precision, fore and co-ordination" "A complex psychomotor skill is characterised by highly organised receptor-effector feedback processes. Thus, a skilled performance, in comparison with one less skilled, is accomplished with (1) less attention to the specific movements, (2) better differentiation of cues. (3) more rapid feedback and correction of movements, (4) greater speed and co-ordination (5) greater stability under a variety of environmental conditions.

The excellent typist, in comparison with the beginner, gives little or conscious attention to hand and finger movements, simultaneously hears or sees more relevant cues to guide sequential movements and ignores the irrelevant, responds more quickly to internal and external sources of information concerning the typing movements, moves rapidly in a rhythmic, unbroken pattern, and types consistently under a variety of environmental circumstances. The level of proficiency achieved by an individual in an activity such as typing indicates the skill he has attained. Thus, we define a skill operationally *as the level of proficiency attained in carrying out sequences of action in a consistent way*"

A skilled performance involves a chain of motor responses which are muscular movements—movements of the arms, legs, fingers and toes. Each movement is an individual stimulus-response (S—R) association. A skilled performance is a chain of such S—R units, hierarchically arranged into response patterns—each S—R unit link acting as a stimulus for the next link. Fitts (1964, p. 245) describes three categories of body movements involved in skilled behaviour. (1) Some skilled behaviour, such as swimming and dancing, involves gross body movements. (2) Some others, as in grasping or manipulating tools with the fingers, involve only parts of the total anatomy. (3) Finally, such skilled performances as driving a car, or playing tennis, or playing on a musical instrument, involves both gross and partial body movements.

The second characteristic of skilled behaviour is that it involves the co-ordination of hand and eye movements. Hence, very often, skilled performances are called *perceptual-motor skills*, to emphasize the co-ordination of perception (the eye) and motor acts (the hand). Swinging the cricket bat or hitting the volley ball requires high degree of hand and eye co-ordination. Vocal skill such as reading involves the co-ordination of eye and speech organs-movements whereas writing skill involves the co-ordination of hand-eye movements.

Skills are chains of S—R units hierarchically organised into complex response patterns. Each complex human skill has a number of S—R units. Piloting an aircraft, driving a motorcar, for example, are very complex human skills having a number of response patterns. The S—R units which constitute the large response patterns are called *subtasks* or *subroutines*. In swimming, for example, the arm strokes, and the leg strokes are the subtasks or subroutines; breathing also is another subtask. As they are hierarchically organised, one should learn particular S—R units and S—R chains before learning others. All the subordinate chains must be learned before you can perform a particular skill.

Classification of Skill

Skills can generally be classified into two broad categories—motor and verbal; and they result in two kinds of learning -motor learning and verbal learning. Some authors use a distinction to separate the fields of motor learning and verbal learning, and discuss each in two different chapters. However, the two areas have much in common with each other and with more complex learning types—the learning of concepts and rules (Fitts, 1964). Hence, some other authors classify skills in different ways from different angles.

How Children Learn the motor Types of Activities", classifies motor activities and skills into three broad categories: (1) *Object-motor Activities* or *Manipulative Skills* include the use of tools in household and industrial pursuits, the use of measuring and recording devices, the use and manipulation of objects used in sports such as balls, and recreation such as swings. (2) *Language-motor Activities or Communicative Skills* include movements of the speech organs (vocal skills, movements of the eyes in reading (reading skills), and the movements in handwriting (writing skills), telegraphy, commercial arts, and mechanical drawing. (3) *Feeling-motor Activities or Affective abilities*—also referred to as manual-aesthetic co-ordinations include such activities as creative dancing and singing, playing of musical instruments and other kinds of fine-arts activities. These activities involve the attitudes and feelings of the individual who practices them, and hence regarded as manual-aesthetic co-ordination.

Merill (1972) thought it useful to classify motor skills based on subject matter taught in schools. He identified five areas of motor skills: (1) *Physical Education and Recreation Skills* include sports, dance and exercise; (2) *Communication Skills* include handwriting, typing and shorthand; (3) *Fine Arts Skills* such as instrument playing, painting, drawing dancing and singing; (4) *Language Skills* have subdivisions such as speech in native language, facial expressions and gestures; (5) *Vocational Skills* include items such as crafts, use of tools, operation of equipments, and operation of machines.

According to Klausmeier and Ripple (*Ibid*; p. 477), skills that are learned in school vary in the amount of motor and perceptual involvement: (1) Skills with *high motor and low perceptual involvement* include walking, dancing, running, swimming, bicycling, and gymnastics. (2) Skills with a considerable amount of *both perceptual and motor involvement* (include playing a musical instrument and taking short hand, (3) skills that *involve low motor and high perceptual characteristic are silent reading,* painting, drawing and the like.

Fitts (1962, 1964) classifies skills on the basis of uniform behaviour patterns. Therefore, they come to be known as *skill constancies.* There are two skill constancies: (1) gross body movements and (2) the extent of external pacing of the activity. *Constancies vary in the level of complexity.* Complexity refers to the varied relationships, of body and objects, at rest and in motion. Fitts (1962, p. 179), describes three levels of complexity in hierarchical order. (1) *The body is at rest* prior to the beginning of the skilled action. Examples of this level of constancy are threading a needle and picking up a pencil. (2) At the second level of constancy, the individual begins the skilled activity either when the *body is in motion* and the *external objects are fixed* or when the *external objects are in motion* and the *body is not in motion* but is set to go. The player in a cricket match swings his bat to hit the pitched ball which is a difficult job. Even in a game of shuttle-cock the same phenomenon occurs. (3) The most complex of the three is when *both the individual and the external objects undergo change just prior to the execution of the skilled activity* as in a game of football or hockey, the players hit the ball in motion while they themselves are in motion.

Factors Affecting Patterns of Action Skills

A number of research findings lead to the conclusion that the particular pattern developed by an individual depends on factors such as (1) maturation, (2) body build, (3) past experience, (4) constitutional characteristics, (5) readiness, which includes motivation and will to learn, and (6) excersise and guidance.

The Cognitive Phase

Skill learning begins with a *cognitive phase*. It is usually of relatively short duration. During this phase the individual does not engage in much practice but does recognise the nature of the skill. In other words, during this phase, the students internalise the skill they are to perform. They develop plans which guide the execution of the skill (Miller et al: 1960). According to Fitts, durging this phase, the instructor and the students try to analyse the skills and to verbalise about what is being learned. "Observing a model, cognising instructions, describing the skill to himself, an learning to make the required responses are typical of this initial phase".

The Fixation or Organising Phase: In this phase the receptor-effect or feedback operations gradually become more highly organised with practice. The behaviour becomes fixed and the skill becomes relatively stable. There is less emphasis on the cognitive aspect and more on the motor. At the first phase, the skill is under voluntary control, only the more obvious cues are differentiated, feedback is elementary, speed and co-ordination are low and the responses are not stable But in the second phase, with practice the skill becomes almost automatic and incorrect responses are reduced almost to zero. The individual learns to organise S—R chains into an overall pattern. Less attention is given to specific actions. Cues that are less obvious are differentiated. Feedback becomes instantaneous. Speed and co-ordination a improve. This stage lasts for days or months. Fitts (1962, p. 188) gives one example. For the typist this phase extends from the point at which he learns the position of the different keys and how the fingers are used in striking them to the point at which he graduates from the first typing course and reduces errors to fewer than one percent.

The Autonomous or the Perfecting Phase: In the autonomous or perfecting phase, there is continuing improvement over a long period of time. As Dececco and Crawford say, "Finally, the autonomous phase is characterised by increasing speed of performance in skills in which it is important to improve accuracy to the point at which errors are very unlikely to occur. In this phase the student also increases his resistance to stress and to the interference of outside activities which he is able to perform at the same time. This is the stage achieved by the expert for whom the performance of the skill has become involuntary, inflexible, and even locked in" (*Ibid*; p. 253). Klausmeier and Ripple comment: "Highly complex skills are performed at an automatic level, The precise combination of motor, cognitive and affective characteristics that produces this stability and continued improvement is not well understood" (*Ibid*; p. 483).

Basic Conditions of Skill Learning: The three basic conditions for the learning of skills are Contiguity, *Practice* and *Feedback*. In skill learning, (a) *contiguity* stands for *timing, co-ordination,* or *proper order*. Two aspects of continuity are important in skill learning: (1) proper sequence of the S—R units and the chains, and (2) the execution of the S—R links in the chains or the chains in the overall response pattern in rapid succession. To illustrate the first aspect: unless these units and chains occur in the proper order we cannot perform the skill. For example, a driver should observe the proper order for stopping a car, first, he should remove his foot from the accelerator and then apply the brakes. You can well imagine the terrible consequences if a novice driver in stopping his car, fails to remove his foot from the accelerator before applying the brakes. Intense training is necessary for an individual to learn the S—R units and the chains in the proper order. Now the question is : "Should the student perform the entire skill or first practice the parts of the skill?" The first procedure is called the whole method, the second is called the part or part-*whole method*. "Using Fitts classification of skills, we conclude that the part method is the better one for teaching those skills in which the constancies are least complex—that is, those in which the body is at rest prior to the beginning of skilled action, as in driving a golf ball or threading a needle. When either body or external objects or both are in motion, the whole method is the better training procedure. We can also conclude that highly coherent, continuous, and complex chains may be learned better by the whole method". In so far as the execution of the S—R links in the chains or chains in the overall response pattern close time succession is concerned, any delay can disrupt the performance of the skill. The S—R nits and chains in skill learning are interdependent. Therefore, the absence of continuity can be seriously disruptive. Timing is of utmost importance in the learning and performance of particular skilled activities.

Practice is a second condition of major importance in the learning of skills. It is the repetition of a response in the presence of a stimulus. "In the learning of skills practice is a way of (1) rehearsing those particular subtasks which are only partially learned; (2) co-ordinating the subtasks so that they are performed in the proper sequence and with appropriate timing; (3) preventing extinction and forgetting of the subtasks; and (4) developing the skill to the autonomous stage of learning" (*Ibid.*) Practice leads to perfect performance. Is it better to practice a task with as little interruption for rest as possible, or is rest beneficial to the learning and performance of skills ? To put it technically: "What are the relative benefits of *massed practice* and *distributed practice*? "Research findings at the present time do not support a simple definitive decision to mass or to distribute practice solely on the basis of categories of verbal or motor learning...in general practice periods should be short and well spaced first, but not so short that the learner will be unable to grasp the act as a unit.

Feedback or *knowledge of results* is the single most important variable in skills learning, John Annet (1964) distinguishes two types of feedback—*intrinsic* and *extrinsic*. *Intrinsic* feedback is the information the student obtains through is own actions. Extrinsic feedback is the information the teacher gives to the student about the effectiveness of his actions. Margaret Rabb (1966) classifies feedback into two types—external and internal. The *external* mode of feedback is information received through the external sensory organs—that is, through vision, hearing touch, smell and taste, Internal feedback is the information obtained from the internal receptive organs, as in the case of kinesthetic feedback. Experimental findings show that feedback acts as a form of prompting, as a form of reinforcement, and also affects motivation as well as learning. Therefore, De Cecco and Crawford (*Ibid*: p. 264) are of the opinion that "(1) Extrinsic and external feedback are important in the early stages of skill learning, while intrinsic and internal feedback are important in later stages; (2) feedback should be immediate or very nearly immediate; giving the student advance information on the performance of a skill may be more efficient than using only feedback for information; and (4) the motivational aspects of feedback may affect performance, or the amount of effort the student will invest in practising the skill".

The Nature and Types of Psychomotor Abilities: It was stated earlier what Fleishman means by psychomotor abilities. For Klausmeier and Ripple, it is "a mediator inferred from the observed consistencies among separate performances". Fleishman and Bartlett (1969) have identified five defining attributes of abilities (1) Abilities are a product of maturation and learning. (2) Abilities are enduring and relatively difficult to change in adults. (3) The present abilities of the individual affect the rate at which he learns new tasks. (4) One ability may transfer to the learning of a greater variety of specific task than another. (5) Abilities are more general and inclusive than skills.

Fleishman (1964) classifies a number of abilities into two broad types: (1) *Physical Proficiency Abilities*, and (2) *Manipulative Abilities*.

Physical Proficiency Abilities: Fleishman has identified

fourteen physical abilities in five broad area: (1) strength–static. Dynamic and trunk: (2) flexibility –speed extent flexibility, dynamic flexibility. (3) balance (grass body equilibrium); (4) body co-ordination; (5) endurance (stamina).

Manipulative Abilities: Fleishman (1964) also reported the perceptual moto abilities or manipulative abilities. They are relatively independent of the physical proficiency abilities and involved in operating machines, manipulating objects and tools necessary in the performance of laboratory skills. The 11 manipulative abilities identified through wide-scale research are : (1) control presicion, (2) multiple co-ordination, (3) response orientation. (4) reaction time, (5) speed of arm movement, (6) rote control, (7) manual dexterity, (8) finger dexterity. (9) arm-hand steadiness, (10) wrist-finger speed, and (11) aiming. As the student moves through the stages of skill learning, the particular combination of abilities which the skill requires changes; the relative importance of abilities varies with improvement in performance (Fleishman & Hempel (1954, 1955). "Any predictions about learning outcomes and any provision for instruction must be cased on the variations in psychomotor abilities that different stages in skill learning require" (De Cecco & Crowford, *Ibid*; p. 271).

Instruction in Skills: The Role of the Teacher, the following steps any teacher should take in teaching skills to his pupils.

Analyse the skill in Terms of the Nature of the Skill, the Learner's Abilities and Developmental Level: The first function of the teacher is to do a task analysis of the skill he plans to teach. He has to analyse the skill in terms of S—R units and chains or in terms of a hierarchy of patterns of chains. This means, the teacher has to analyse the skill into *subtasks* or *component skills*. To analyse the skill in terms of constancies, coherence, continuity, and complexity is also a genuine attempt. The most thorough task analysis combines the use of each of these methods. While analysing the skill, the teacher should not forget to take into account the abilities of the learner in the present and his level of development. This leads to the second step the teacher should take in teaching a skill.

Assess the Entering Behaviour of the Student. Entering behaviour describes the present status of the student's knowledge and skill in reference to a future status the teacher wants him to attain. Therefore, the second function of the teacher is to determine the adequacy of the student's preparation for learning the skill. Therefore, he has to assess the entering behaviour of the student. Here again he is presented with a choice of analytical methods. He can assess entering behaviour in terms of subtasks or of psychomotor abilities. Once he determines the existent psychomotor abilities in the learner, the teacher must make an analysis of the abilities required by the skill and must match student ability and task demands.

Arrange for Training in the Component Units, Skills or Abilities. This step is needed to provide the student with the opportunity to learn missing S—R links or component skills or to develop the prerequisite psychomotor abilities; and also to provide the student with the opportunity to learn the skill components so well that he can focus his attention on the new aspects of the complex task he is learning.

Demonstrate and Describe the Skill for the Student. The next function of the teacher is to demonstrate the correct responses to be made in learning the skill and to guide initial responses verbally and physically. According to Bandura and Walters (1963) new behaviour units, or chunks of behaviour, are learned initially through observing and imitating a model rather than being shaped through successive approximations involving differential reinforcement. A good demonstration by the teacher provides an overview of the skill to be acquired and a model to be imitated. If a demonstration is to be effective, the teacher must be able to perform the total skill and its various components well. Information given verbally as well as in a live demonstration may be useful for describing the nature of the final performance desired. Miller et al. (1960), have shown that verbal instructions can be used successfully to outline a strategy or method of attack for developing a skill. According to Fitts (1964, p. 260), in skill learning the student usually listens to instructions, observes demonstrations, tries out different component skills which he already has available, and then somehow starts learning the new skill. The purpose of the fourth step is to help the student develop a plan for the execution of the skill to be learned. Therefore, much rehearsal is needed. At this stage, the teacher should emphsise only the essential characteristics of the performance. Motion pictures are especially useful in demonstrating movement and timing. Demonstration should eventually provide for verbalisation and subsequently for student practice of the skill.

Provide for the Three Basic Learning Conditions: Contiguity, Practice and Feedback. In teaching skills, the three basic learning conditions - contiguity, practice and feedback—should be combined into a single step because the teacher must provide them concurrently rather than sequentially. His success in teaching a skill will depend not only on how well he provides for each of these learning conditions but also on how skillfully he can combine them into a single teaching situation.

The teacher, *to provide for contiguity*, must teach the student proper co-ordination and timing. "For proper co-ordination the student must learn the proper order, or sequence, of the units or subtasks of the skill, For proper timing, the adjacent units in a chain or adjacent subtasks in a hierarchy must occur simultaneously, without delay and unnecessary pauses. Previously we discussed three methods for providing continuity: the whole method, the part method, and the reverse part method of Gilbert.

Practice, of course, is essential to the improvement of skills. The practice must be carried out under desirable conditions, however, Distributed practice is more useful than massed practice in the learning of psychomotor skills. According to Klausmeier and Ripple (*Ibid*; p. 509), "Whole part arrangements, the context in which the skills are practised, and the length and spacing of practice sessions are the more important conditions associated with effective practice." The teacher should adhere to these principles to *provide for practice*.

Finally, the teacher has the responsibility of providing feedback. Knowledge of results is one of the most powerful variables in skill learning. There is no improvement without a knowledge of results; there is progressive improvement with such knowledge; there is deterioration after its withdrawal. Obviously, feedback of information is required to give the student knowledge of results. "The need for extrinsic and external feedback in the early phase of skill learning imposes this immediate responsibility"

"Information can be given verbally in such simple statements as "correct" or "incorrect"; a verbal analysis can be given of anything the student has produced; and the results of performance tests of all sorts can be incorporated in charts or given to the student directly. Verbal presentations, however, do have limitations. In some skills the actions cannot be explained verbally, nor can the accuracy be measured. In this case the teacher may demonstrate and the student, by observing, can compare his performance with the teacher's ...the learner can also be encouraged to engage in mental practice of the task both before and after he has completed it. This mental practicing is sometimes useful in helping the learner to internalise his knowledge of results both in ascertaining what he has done well, and, possibly, where he has done an error"

14

LEARNING OF CONCEPTS AND PRINCIPLES

A concept is made, not born, individuals do not agreed on the nature of concepts. Therefore, it becomes difficult for teachers, curriculum experts and designers of instructional material in identifying the concepts that students may learn at successively higher levels. The problem becomes all the more difficult when the nature of learning concepts and the procedures to teach them are taken into consideration. *Principles* too are important outcomes of learning. *Relationships between concepts establish principles.* One example is : "Winged creatures fly". Principles are vital to solving problems. If the individual has learned the needed concepts, he has to learn relationships among concepts so that he can learn how the principles are formed. The learning of concepts themselves also involves the learning of relationships—relationships among attributes and rules. This shall become apparent later in this section of this chapter. It is believed that the internal and external conditions, essential for learning either concepts or principles, are much alike. It seems logical, therefore, to concentrate on the learning of concepts and principles at this place.

The Nature of Concepts

Some definitions of a concept have already been given

earlier. The S—R psychologists and the cognitive psychologists have their own learnings in defining a concept. Without entering into complications, let us begin with some relatively simple definitions of a concept. We may say: "Concepts represent abstracted characteristics of many events, stimuli or symbols". Also we may say "A concept is a class of stimuli which have common characteristics. These stimuli are objects, events or persons'. Though somewhat sophisticated, a very clear definition of a concept is: "A concept may be defined as a way of grouping an array of objects or events in terms of certain characteristics that distinguish this group from other objects or events. By grouping together things or events that have common characteristics one in able to (1) reduce the complexity of the situation. (2) identify more clearly the objects or events, (3) make more meaningful the learning experience, and (4) relate objects and events in the special grouping" (Symonds, "What Education Has to Learn from Psychology. VI. Emotions and Learning," 1958, quoted by Garrison et al; *Ibid*; p. 252). A concept is to a stimulus but a class of stimuli. Thus, book, day, student, beautiful girls are concepts. Proper names such as Miss Leela. The Taj Mahal. Tapas Girl, Shakespeare's *Hamlet*, the Iran-Iraq War. Are not concepts. These are particular (not cases of) stimuli, that is objects, events, or persons.

We form concepts in our day-to-day life by continually coming across members of the particular concept cases and deciding that they are all linked by unifying attributes. Thus, concepts are made, not born. To make you remember one other thing: it was stated earlier that an individual's concepts change with increasing experience with instances of the concept, including additional verbal information about the concept; that there occurs elaboration of concepts with age. Each individual has his own concepts, based on his nonverbal and verbal experiences.

The individual's connotative meaning of a word corresponds to that characteristic of a concept called *psychological meaning*- fullness which changes with increasing age. To illustrate: Consider your own concepts of "*justice*", "*courage*", "*pleasure*, "*space*", "*time*", *democracy*", and "*secularism*" now and compare them with those you had when you were a preadolescent. Thus, psychological meaningfulness may be treated as one of the characteristics of a concept.

A second characteristic of concept is *structure*. Bruner. Goodrow, and Austin (1977) see any concept as having five *elements*: (1) name, (2) examples (positive and negative), (3) attributes (essential and nonessential), (4) attribute values, and (5) rules. Understanding a concept means knowing all the elements of the concept, that is, its *structural components*.

The name is the term given to a *category; Green fruit, black dog, lovely girl, blue circle, cold courage, dirty ghetto, surgical operation,* are all *names* given to a class of experiences, objects, persons, figurations or processes. The items commonly grouped together in a single category may differ from one another in certain respects; dogs, for example, vary greatly—from a mastiff to a poodle; yet, the common features cause them to be referred to by the same general term. If students, know a concept, they can easily learn the name for it, and their verbal expressions will be more articulate.

The second element, *examples*, refers to instances of the concept. The instances of concepts may vary from one to an infinite number. There are concepts for which there is one instance, for example, *galaxy, the earth, the Pacific ocean, the moon, the pyramids*. Other concepts are there which have many instances, all of which, however, are identical, Examples are: *inch*, for measuring length, *ounce* for measuring weight, punctuation marks such as *comma*, and similar units in other fields. On the other hand, the instances of many concepts are infinite, for example, the instances of numerals are infinite, also it appears, *grains* of *sand* and *drops* of water have an infinite number of instances. Instances also vary according to openness to sensory experiencing. "*Apple*" is available to sight, touch, smell and taste. But *"atom"* and *"genetic code"* are not, and therefore, very difficult to represent Finally, instances may be positive or negative. There are also noninstances too. Positive instances possess the attributes of the concept. A *negative* instance has some of the same attributes as the positive instance, but not others, Suppose there is an array of fruit in front of you, mostly a variety of oranges, but one or two apples, and a pear. Each item is an example or instance. If our concept is *"orange"*, the apples and pears are negtive examples, and the organges are positive. A rectangle is a positive instance of *quadrilateral* but a negative instance of triangle. It is a noninstance of *mammal* and of all other concepts that are not geometric forms.

The third and fourth elements of a concept are *attributes* and *attribute values*. An *attribute* is a distinctive feature of a concept and thus varies from concept to concept. Attributes are the common features or characteristics that cause us to place examples in the same category. *"Green squares"* is a laboratory concept. This concept has two attributes: *Colour* and *form* or *shape*. Colour, of course, can vary from concept to concept and, therefore, qualifies as an attribute. As we can have "green circles", and "green triangles", so we can have "green shirts" and "green bananas". In the case of a fruit, the attributes are: fleshy or dry, seed enclosing, sweet or sour, and edible or not edible. We can attach a value to an attribute. *Attribute values* are the particular variations an attribute may undergo. If colour is an attribute, it may have several values: red, green, yellow, blue, pink, brown. Similarly, shape may have several values: round, triangular, square, elliptical, rectangular and octogonal. Concepts vary in the number of values their attributes have. The concept *"human being"* has two values: man or woman, dead or alived, and married or single. Other concepts may have attributes with a range of values. The colour of the homely apple varies from green to yellow to red, and within this range many different colour combinations are acceptable for apple. We speak of the range of any given dimension as its value range. Blue or black or purple is outside the acceptable *value range* of colour for apple. When an attribute has a wide range of values, we can use other attributes to identify the concept in question. In identifying an *orange*, for example, in addition to colour, we also observe the attributes of shape, size, and texture. Similarly, in identifying a *dog*, in addition to colour you can take into account the attributes of shape, size, texture and behaviour.

Attributes of concepts vary in *number, dominance, relevance,* and *discriminability. Blue squares* has only two attributes—colour and form. *Small red circles* has three attributes—size, colour, from. A *dog* has five attributes colour, shape, size, texture and behaviour. *Democracy*, Secularism, and *Human Rights* have volumes of attributes The four relevant attributes of *quadrilateral* are closed figure, plane figure, simple figure, and four sides. In defining *square*, the same four attributes and two others—equal length of sides and equal size of angles—are used. Thus, the equal length of sides and equal size of angles are *relevant* for *square* but not for *guadrilateral*. The size of the figure, the position of the figure, and the way in which it is represented are *irrelevant* for identifying instances of both squares and quadrilaterals.

Dominance refers to the fact that some attributes are more obvious than others. For example, colour form concepts such as *blue squares* are more dominant than number—colour concepts such as *one blue*. Thus dominance refers to the concept as well as to its attributes. Dominant concepts have dominant attributes. Teachers must give aural or visual emphasis to attributes which are obscure and yet important in identifying the concept. *Discriminability* refers to the extent to which the attributes are attended to and cognised. Thus *dominance* and *discriminability* are related characteristics. Attributes can be made discriminable by various means. Size can be made more discriminable by using both large and small examples such as *elephant* and *mouse*. By holding *size* as constant as possible, and using heavy lines to represent the sides discrimination of the *number* of sides in triangles and quadrilaterals can be facilitated; The teacher can also tell his pupils what to look for. Teachers should remember that as the discriminability of the attributes decreases, concept attainment becomes more difficult.

Types of Concepts

Different rules for joining attributes lead us to different types of concepts. Attributes combine in three different ways to produce three types of concepts: *Conjunctive concepts, disjunctive concepts, and relational concepts.* Bruner, Goodnow & Austin (1977) illustrate attributes of concepts represented by the words *red, mammal, older,* and *strike* in the following manner. (1) *Red* is a simple affirmative concept comprised of one property or dimension (2) An animal that simultaneously, or *conjunctively,* has three attributes—warm—blooded, mammary gland, bearing live offspring -is classified as a *mammal* "In a *conjunctive concept* the appropriate values of several attributes are jointly present" (De Cecco & Crawford, *Ibid;* p. 291). Three dark gold circles is a conjunctive concept. it has three attributes—*number, colour* and *form.* Each attribute has a particular value, respectively, three, dark glod and circle. Another conjunctive concept is dog; it is different from cats, cows and horses. To produce a conjunctive concept, attributes and values are added together. Conjunctive concepts are often the easiest to learn and to teach. (3) A five year old child is *older* than one aged four, but younger than a six-year-old; older is a *relational concept.* "A *relational concept* is one that has a specifiable relationship between attributes" (De Cecco & Crawford, *Ibid;* p. 292). *Distance* and *direction* are relational concepts, *Distance* specifies the relationship between two points; it refers to the separation of these points. *Direction* also specifies a relationship between two or more points; it refers to the movement from one to another point. Carroll (1964) treats mass, weight, time, average longitude, many and few as relational concepts. Relational concepts are more difficult to learn than conjunctive concepts. Both the concept *distance* and the concept *direction* have as their attributes points in space and time. What distinguishes them is the difference in the relationship of the same attributes. Thus, the children as well as adults get confused. (4) A *strike* in baseball represents a concept where attributes are joined by a *disjunctive rule*—and /or a strike may be a ball thrown in the strike zone and called by the umpire or it may be a pitch swung at and missed or it may be a foul. "A *disjunctive concept* is one that can be defined in a number of different ways. In such concepts attributes and values are substituted for one another. *Two figures and/or two circles* is a disjunctive concept" De Cecce & Crawford, *Ibid*; p. 292). A much more difficult disjunctive concept is *five dark red squares* or *any constituent thereof*. Disjunctive concepts are often difficult to learn because the seemingly arbitrary equivalence of their attributes. As teachers we must invest greater effort and ingenuity in teaching them.

Frayer, Frederick, and Klausmeier (1969) have added *another element* to the five we have already discussed. To this element they name *Hierarchical Patterning*. Man fits into a classification system—*vertebrate, mammal, primate, Hominidae. Homo sapiens*. Man is the only living survivor of the family Haminidae. "Note than as one goes down the list in this order each concept increases in number of defining attributes. In other words, each concept successively is defined by the relevant attributes in the higher-order concept and also by the additional ones required to differentiate among the various groups in the same overall classification".

In addition to psychological meaningfulness and structure concept has a third characteristic—*Transferability*. "Once a concept has been acquired, there are four possibilities for use in, or transfer to, other situations (1) One categorises other instances, when first encountered, as belonging to the concept; also one can categorise other instances when first encountered as not belonging to the concept. (2) One can more readily cognise other concepts as superordinate, co-ordinate, or subordinate, (3) One can use the concept informing and understanding a principle and in solving problems where the concept applicable. (4) The learning of one concept facilities the learning of other concepts".

The Nature of Principles. A principles is defined by Gagne as a relationship between two or more concepts. According to him, concepts are essential for the cognising of principles and the solution of problems. "A problem is encountered when one does not have a solution, a method, or both to deal with a situation that one must resolve" Principles are sometimes called rules or generalisations. This was stated earlier and some examples where given. To recapitulate one or two examples : (1) All men are created equal (2) The negative sentence is made by placing the adverb not after the anomalous finite. Thus 1 am → I am *not*. (3) Six multiplied by the three equals eighteen. These statements are not principles. (1) Ashok likes Ramesh. (2) Evergreen trees are found in India. (3) The Congress won the last election. The first set of statements qualifies as principles because the sentences state relationships among concepts. The second set of statements does not. For example, in the sentence "*Ashok likes Ramesh*"; *Ashok* and *Ramesh* are not concepts. Likes is a concept but it does not establish any relationship between concepts. Principles may be treated as "*If – then statements*", examples are (1) If an object has six sides, then it is three dimensional. (2) If the subject performs the action of the verb, then you use the active voice. (of – then statements indicate the proper ordering of the component concepts. The proper arrangement of the concepts results in statisfactory learning of principles.

(1) concepts reduce the complexity of the environment. (2) Concepts help us to identify the objects of the world around us, (3) concepts and principles reduce the necessity of contain learning. (4) Concepts and principles provide direction for instrumental activity; we know in advance the

action we can take; placing the object or person in the right class enables us to arrive at important decisions, and thus helps us for problem solving (5) concepts and principles make instruction possible. (6) Frederick Mc Donald (1965) defined a stereotype as "a rigid or inaccurate concept imperious to experience". Concepts can be stereotypes. Therefore, the teacher must sometimes provide corrective experience for an additional use of concepts, stereotypes. As concepts, stereotypes can be changed, sometimes, when the student is provided with a wider range of positive and negative examples than those which he has previously experienced.

Teaching and Learning of Concepts

Securing information is essential for learning concepts. Information about concepts and their applications may be acquired through reading and studying printed material. "Factual information is information that is discriminated by individuals who share the same cultural background and is also accepted by them as correct or appropriate....This kind of information is essential to learning to speak, to read and to write and also informing concepts and in problem solving'. Concept and principle learning is, therefore, closely related to language learning. Words represent concepts. Much of concept learning occurs through the use of words.

Carroll summarised the complex relationships among words, meanings and concepts which indicate the closeness of relationship between language learning and learning of concepts and principles. According to Carroll, the "words" in a language are a series of physical entities—either spoken or written. A "meaning" is a standard of communicative behaviour that is shared by those who speak a language. "Concepts" are classes of experience formed in individuals either independent of language processes or in close dependence on language processes. The interrelationships found among these three series—words, meanings and concepts—are complex.

A "word" may have many "meanings", and a "meaning" may correspond to several different "words". There are societally-standardised "meanings' and individually formed "concepts". The relationships between these two sets are of a very complex nature. It is a question of how well an individual has learned these relationships. In the sphere of language and concept, education is largely a process whereby the individual learns either to use societally-standardised words and meanings to the concepts he has already formed, or to form new concepts that properly correspond to societally standardised words and meanings. A "meaning" of a "word" is therefore, "a societally-standardised concepts", and when we say that a 'word" stands for or names a "concept" it is understood that we are speaking of "concepts" that are shared among the members of a speech community.

On the relationship of parts of speech to concepts, Carroll is of the opinion that many words or higher units of the linguistic system come to stand for, or name the concepts that have been learned preverbally, for example, proper and common nouns; adjectives; and verbs of action, perception and feeling; in other words, particular things or classes of things, qualities and events. Prepositions like *in, to, above, below, beside,* and *near* correspond to concepts of relative spatial position. Conjunctions like and, *but, however,* and or correspond to a concepts of logical inclusion and exclusion, similarity and difference of proportions, etc.

Referring to R.M. Travers (1967), Gibson States: "The availability of language facilitates the task of concept learning considerably, because language services as an economical and convenient way of ordering and classifying many different sets of observations. Language also contributes to greater uniformity in the use of concepts by different people. In addition, children are constantly encountering new words that serve as cues to the existence of concepts they don't yet know about, but perhaps should. Their natural inquisitiveness leads them not only to learn the meaning of the word, but also to place that word in a conceptual category. Finally, language enables the learner to grasp the relationship among different concepts and to express these relationships in a meaningful, precise, and understandable way" (Gibson, 1976, p. 244).

Keeping in view the discussions carried on so far, let us examine the role of the teacher. The first question the teacher should ask and answer is: *What is the performance expected of the student after he has learned the concepts*?" Commonsense warrants that the answer is: "In concept learning, the expected performance is the correct identification of new examples of the concept. For example, if the teacher gives a number of English sentences and identities *direct object* in the first sentence underling it, and asks the pupils to underline similar words in the other sentences, they should be able to do so. To illustrate: "They were throwing stones)' (Subject Verb Direct Object). (2) I dug a hole. (3) He cut his finger. (4) He does to like cold weather. (5) She language a merry laugh.

The next question is : "*What concepts should I teach first, and what next, and what next*?" In other words, "how are you to choose the concepts you wish to teach?" You have already learned the nature of concepts. Some are simple and concrete while some others complex and abstract. The former kind should be taught first before you go to the latter type. You should teach your pupils about bags, balloons and balls before you begin to teach courage, cruelty or chivalry. You should put to pedagogical use what you learned about attributes, their number values, dominance and relationship with other attributes. The number of attributes of complex concepts should be reduced. Dominant attributes should be taken into account. "Two general procedures reduce the number of attributes of complex concepts: you can ignore some of the attributes and focus on those you think most important, or you can code the attributes into fewer patterns".

How is the teacher to know how much his pupils already know about concepts before he begins to teach about concepts to beginners? The answer to this questions depends on the entering behaviour, that is, the present status of the students knowledge and skill in reference to a future status the teacher wants him to attain and the stage of cognitive development the student is in, as described by Piaget. Therefore, one of the fundamental tasks of the teacher is to establish the verbal association, if the pupils have not already learned, necessary for learning the concept. We are beginning to realise that verbal learning and concept learning have much in common. Let us use the concept *dog* to illustrate this point. The student must be able to respond "*dog*" when the teacher shows the picture of three different kinds—a bulldog, a greyhound, and a pekinese. The responses are what enable the teacher to ascertain whether the student has the appropriate verbal response available.

The teacher should emphasize the attributes of the concept. The reason for this being: attending to likeness and differences among things, qualities, and events is essential to subsequent classification.

As acquiring the names of concepts, attributes and instances facilitates the initial learning of concepts, the *teacher should establish the correct terminology for concepts, attributes and instances.*

The teacher should indicate the nature of the concepts to be learned, because cognising the definitional basis and the structure of concepts to be learned facilitate concept learning. The implication of this guide is that when a unit of study is first started, the teacher can take steps to assure that the defining properties of the concepts to be learned will be emphasised, that the necessary terminology can be spoken and read, and that students understand the nature of the concepts that will be learned.

Cognising the attributes or other properties and the rules that define the concept is facilitated through encounters with positive and negative instances of the concept. Therefore *not only* should the teacher provide positive and negative examples of the concept but also present the examples in close succession or simultaneously. Janellen Huttenlocher (1962) suggests that a combination of positive and negative instances of the concepts rather than all positive or all negative examples, produces a more effective learning. Enough positive examples to represent the range of attributes and attribute values of the concept and enough negative examples to eliminate irrelevant attributes which students are likely to include as part of the concept, should be presented to the pupils. Travers and his associates ("Research and Theory Related to Audiovisual, Information Transmission, 1964) prove that drawing or animated moving pictures which show only the important attributes of the concept are considerably more effective than direct experience. Marks & Remond (1951, p. 428) state: "We know that people do better in school problems than in real-life problems. Perhaps the academic halls have an atmosphere conducive to analysis. An almost simultaneous presentation of the examples of the concept is more effective than a successive presentation of instances, according to Cahill & Hovland (1960), and Kates & Yudin (1964). Presenting several instances of the concept simultaneously permits discrimination of the attributes more readily and also reduces the memory load. However, successive presentation is probably the most common teaching practice because our teaching habits sometimes prompt us to focus the students attention on one thing at a time.

Encourage and guide student discovery because inferring a concept inductively or deductively requires cognising the defining attributes and rules, remembering information, and evaluating information. As Suchman (1964) suggests: (1) present students with a problem that is real and meaningful; (2) encourage and guide students in gathering information; (3) provide a responsive environment in which students get accurate feedback promptly so that they can ascertain the adequacy of their responses. The research finding of

Klausmeier and Mainke (1968) shows that guiding the discovery process by giving students some information about the structure of the subject matter and about a strategy or a principle to use in securing and evaluating information from positive and negative instances resulted in better concept learning.

Provide for use of the concept to help the learner cognise other instances to the concept: cognise other concepts as co-ordinate, subordinate and superordinate; to use the concepts in forming principles and in solving problems so as to extend the learner's concept.

Provide occasions for student responses and the reinforcement of these responses. Students response and reinforcement of these responses are crucial learning conditions for concept learning. The primary purpose of reinforcement is to provide informational feedback to the student on the correctness of his responses. Reinforcement has motivational aspects for which it is so very important. The teacher should focus on the student's responses - not on the student. The mode of the response should not be shifted, for example from spoken to written or vice versa and from writing to drawing or drawing to writing.

Assess the learning of the concept. This step emphasises generalisation, or the ability of the student to make the conceptual response to a new but similar pattern of stimuli. In teaching the concept *dog*, for example, you might have shown pictures of Alsatians, Saint Bernards, German shepherds, and English Bulls. In this step then you present a picture of a poodle, or terrier, or a sad-faced cocker spaniel. If the student is able to identify the new example of a dog, he has learned the concept.

Encourage independent evaluation of the attained concept; but of course, when the pupils are ready for it. The principle that supports this step is that evaluating the adequacy of one's concepts is essential for independence in concept learning.

Problem-Solving and Creativity

We are in the atomic age; rather, the space age, problems, on cannot count how many, and great or small, confront us everyday in this modern world of increasing complexity. We are in a mortal struggle for the survival of our way of life in the world. One of the aspects of this struggle is military preparedness. Besides this, scientific and cultural as well as economic and political problems challenge us and demand solutions. These are great problems. We face small problems too. According to Getzels (1964) there are two types of problems: *presented* and *discovered*. In school or out, most of the problems we confront daily are not of our contriving. They are *presented* to us. Some of these are simple, but others complex. The *presented problems* are sharply defined and easily recognisable. To illustrate a simple problem: In our country, indiscriminate deforestation has resulted in the price of fuel-wood and charcoal rising high. An underpaid employee and his wife feel like buying a gas stove. But they do not have the required cash. What can the employee do? (1) He can borrow some money from his friend and buy the article. (2) Promising the shopkeeper that he would pay the cost on instalmental basis he can bring the article come. (3) He can ask for a loan from his Provident Fund Accumulation and buy the stove. This is a simple problem. The employee can solve it. A presented problem may be as complex and digging a canal from Brahmaputra to Kaveri to control floods and for irrigation purpose. Be they simple or complex, the solutions to such problems also are sharply defined and easily recognisable. The problem and the goal always remain clear. There is a standard way or ways—to solve it.

But what is a problem? "In general it is any conflict between one stimulus situation and another one we want to produce—our goal" (Morgan & King, 1975, p. 181). According to Owen and his co-authors, "...a *problem* consists of a situation requiring a solution in which one does not have a ready, immediate answer...Some psychologists would add to the definition "and in which the person of animal involved is motivated to find a solution". In short, if a solution is not demanded, no problem exists" (1978, p. 269). Very true. A problem is not a problem just because somebody says it is. It is the person who has to solve the problem; who must see it as a problem. Klausmeier and Ripple also say the same thing using different terminology. "An individual is confronted with a problem when he encounters a situation where he must respond but does not have immediately available the information concepts, principles or methods to arrive at a solution" (*Ibid*; p. 439).

School children encounter a number of presented problems. One boy has been elected the captain of the football team. He has to organise the team excellently well to prove his competence. A second boy desires to top the class in all the school subjects; but his parents are too poor to provide him all the books he needs. A third boy encounters are really difficult sum in algebra and he does not know the needed methods to work it out. A fourth boy has run out of cash and fears he might not be able to get home for the holidays. He wrote a letter to Dad. His Dad has not yet responded. A fifth boy has a chemistry examination next Wednesday. He has not yet been able to read his textbook and decipher his notes. These children are motivated to reach their respective goals. They have not reached their goals yet They encounter problems which must be solved. Such problems get thinking started.

In contrast to *presented problems*, there are *discovered problems*. Discovered problems are those which people have sought out. They are neither sharply defined nor easily recognisable. We are generally concerned, as teachers, with presented problems. For each problem, there is one "best" solution. But not so in the case of discovered problems. The discovered problem is one that most people do not see. For the first time Newton asked the question: "Why does an apple fall to the ground. "But why should we bother about such things? Don't we have enough presented problems in hand demanding solutions ? Why should teachers inspire students to look for problems and instruct them in how to find them? Half-a-century ago Einstein answered such questions. He said: "The formulation of a problem is often moe essential than its solution, which may be merely a matter of mathematical or experimental skill. The wheel, the steam engine, the aircraft, printing and penicillin, robots and computers, quantum theory and quantitative psychology, the teachings of Buddha and Jesus Christ—are all derived from discovered problems. Discovered problems generate a train of thought in the individuals who have sought them out. Creativity, problem-finding and problem-solving are three concepts inextricably bound. Creative imagination is needed to find out a problem; and once you find out a problem, you must try for its solution.

In the earlier chapters it was stated both simplicity and explicitly that learning is a form of problem solving activity and problem solving is a form of learning. These two mean the same thing. If so, what relevance does the notion of learning problem-solving activity or problem-solving skill have here? One noted experiment in problem solving by George Katona, compared groups of students who had learned to solve problems by two different methods. Students falling in the first group memorised the solution to a problem which required the moving of lines to make a new geometrical shape. Students falling in the second group spend the same amount of time as the first in working on several such problems; the solutions to which were explained. Retests of both groups on similar problems showed a decided superiority for the second group which really understood the nature of the problems in commenting about this and like experiments, the investigator wrote: "Pupils should learn to learn—that is the best the school can do for them. They should not merely learn to memorise—they should learn to learn by understanding". The answer to the question we have raised is implicit in the words of Katona.

How Do We Solve Problems?

The connectionists and the cognitivists differ in their respective approaches to problem-solving. Let us examine the connectionist view, first E.L. Thorndike held that problem-solving was essentially a process of trail and error. His cats solved their problems by trial and error. No insight was involved. Skinner offers another behaviourist explanation: "A person has a problem when some condition will be reinforcing but he lacks a response that will product it. He will solve the problem when he emits such a response" (1974, p. III). But how does he emit the response? Perhaps, by changing the environment, which produces the stimuli that control behaviour.

Hull stands unique among the behaviourists. He developed the principle of "*The Habit – Family Hierarchy*" which makes possible the explanation of response equivalences and other appropriate reactions in novel or problematic situations, such as those found in reasoning experiments This principle of Hull which he calls an *intermediate mechanism* is used as a theoretical tool to explain a variety of behaviour. "Hull believed in association of stimuli to responses, and he used *habit strength* to denote the strength of a given association. Learning in Hulls theory consists of the growth of habit over a series of reinforced trials". In a natural environment there is the organism, a starting point and a goal. There are multiple routes – from the longest to the shortest – between the starting point and the goal, "the organism learns alternative ways of moving from a common starting point to a common goal position where it finds need satisfaction. These alternatives constitute a "family" an equivalent responses, called a habit family, because of an inferred integrating mechanism." The starting responses of longer routes are more remote from reinforcement than the starting responses of shorter routes.

Hence the shorter routes are more strongly reinforced and more strongly conditioned towards the anticipatory goal. "As a consequence, the alternative behaviour patterns are arranged in a preferred order. The less favoured routes are chosen only, when, the more favoured are blocked. It is this set of alternative habits, integrated by a common goal stimulus, and arranged in preferential order, that constitutes a *habit-family hierarchy*. It was further deduced by Hull that if one member of a habit-family hierarchy is reinforced in a new situation, all other members of the family share at once in the tendency to be evoked as reactions in that situation (Hull, 1937). This makes possible the explanation of response equivalences and other appropriate reactions in novel or problematic situations, such as those found in reasoning experiments" (Bower & Hilgard, pp. 98-105).

To simplify: Hull believed, every individual has a behaviour repertoire of possible responses, for any given drive or need. He has a whole family of possible responses. But the responses are not of equal strength to be exercised at random. They are arranged in a hierarchy, with the one at the top, the most reinforced, and the one at the bottom, the least. The top most response is the one that has provided the greatest reinforcement in the past. When the individual has a problem, his first response will be from the top of the relevant hierarchy. But that may not solve the problem. So he works his way down the hierarchical ladder until he finds a response that does solve the problem.

To illustrate : Tomorrow is your examination on Educational Psychology. It is midnight. Sitting in your hostel room, you are deeply engaged in reading this text. Suddenly you feel, you are thirsty—really hasty. You look around for the water jar. You broke the earthen jar in the morning. There is a water pipe downstairs. Suddenly you remember, it has also one out of order. You look across the street for the shop where cool-drink is sold. To your dismay, you find the shop closed. There is a cafeteria about one kilometer from your hostel where coffee, tea and snacks too are available. But the cost is high and the stuff bad. Beside the train station, about 3 kilometers from your hostel, there is a round-the-clock cafeteria where coffee, tea, chocolate milk shake and snacks are available. The price is reasonable; the preparation excellent. You do not have a bicycle and all your friends are gone to bed. To go in a rickshaw will cost you too much.

Here is a problem for you to try

Two train stations—B to the east of A—are 100 miles apart. At 9. A.M. one rainy Tuesday morning, a train leaves each station heading towards the other. The train going east speeds along at 60 miles an hour. The westbound train travels at a rate of 40 miles an hour. Just as the eastbound train pulls out of the station, a crow leaps from the engine and flies ahead to the oncoming westbound train. When the bird reaches the other train he turns around and immediately flies back westward without losing any speed or time in turning. Upon reaching the eastbound train, it turns around and heads back to the other train again. The crow continues to fly, back and forth between the two trains at 80 miles per hour as the trains approach each other. How many miles will the bird fly until the two trains meet each other?

Stop reading further and solve the problem in two minutes; after which read further whether you succeeded or failed in solving the problem. This is what you might have done. You would have tried to figure out how long it takes the crow to fly from the east bound to the west bound train and vice versa. While it is possible to solve the problem this way, it helps if you are good at mathematics, have paper and pencil, and may be even are handy with a small electronic calculator. By this procedure, you might have solved the problem in two minutes; you might have failed.

An intelligent guess, an insight into the fundamental nature of the problem could have helped you to solve the problem in two minutes, without giving any one an inkling of doubt of your success. If you are a Gestalt psychologist—a person belonging to the school of psychology that emphasizes the importance of perceptual organisation—you would have argued for a different approach to the problem. You would have rearranged the problem so as to gain an insight into its fundamental nature. What about reorganising the problem around the time it takes the train to meet? Since they are one hundred miles apart, they will meet when the east bound train travels 60 miles and the west bound train 40 miles. This will take one hour since the east bound train travels at 60 mph and the west bound train speeds along at 40 mph. Now, how far can the crow fly in one hour? The answer is easy: 80 miles. The problem is solved with little fuss or bother. The trick was seeing the problem in a productive manner.

Gestalt psychologists, the forerunners of the present day cognitivists argue that problems are efficiently solved when organised into a good structure. They say that a problem exists when a person's perception of the world does not provide a solution, and that to find the solution the person must reorganize his or her perceptual field. The cognitivist theories of problem-solving emphasize the concept of insight, defined as the sudden perception of the relationship in a total situation, that is, the relationships between the organism, the goal, and the intervening obstacles. It is also referred to as a restructuring of the perceptual field" suddenly the problem snaps into focus and the solution is then obvious. Remembers Sultan Kohler's chimpanzee? With a flash of insight, Sultan realised that the short stick could be used to pull in a longer stick that would reach the bananas. In humans, this insight is often accompanied by some verbal exclamation like "ah ha! Now! See how to do it." It was a similar flash of insight, no doubt, that inspired Archimedes to run naked from his bath and race through the streets of his native Syracuse in Sicily yelling "Eureka, eureka!" ("I have found it") "It" was the solution to the problem put to him by his friend King Hieron, and you know the whole story that has come down to us and along with it the famous Archimedes Principle, "Thus, in many text books we come across the terms "Eureka !" experiences, and "Gestalt concept of ah ha!", used to explain the concept of "insight" in learning and problem-solving.

The cognitivist approach is very different from the connectionist view, in which learning is considered largely a trial-and-error process. For the Gestalt psychologists, *productive thinking* is involved in learning and problem solving. *Productive thinking* requires producing a new organisation. "Thus, while associationists talk about the function of petition for building habits, Gestalt psychologists argue that repetition by itself is useless. Understanding—not the formation of mere associations between stimulus and response—is "the key to problem-solving".

But now, things are not what they were, "Modern Cognitive psychologist believe that it takes more than reorganisations of the perceptual field to explain problem-solving, and they are interested in how people acquire, classify, and use information to arrive at answers. One approach to problem-solving, often referred to as hypothesis—testing, involves intelligent guesswork. You perceive a problem. Using information stored in your memory or newly gathered, you propose a tentative solution, or hypothesis. Then you try out the hypothesis. If it stands up under the test, you have solved the problem. If the hypothesis does not stand up, you revise it or discard it and develop another hypothesis".

This explanation differs from Hull's. His theory assumes the problem-solver's control by the environment. Hypothesis—testing is based on the problem-solver's collection of information and the application of that information to the situation.

Gagne (1964) believes problem-solving, sometimes called rule learning, to be closely related to principles learning since it combines two or more previously learned principles or rules into a higher-order principle. And like concept and principle-learning, problem-solving represents acquisition of substantive knowledge. "Gagne" (1964) uses this example. A student beginning algebra is given a problem—multiply x^2 and x^3 -which he has not seen before. To solve this problem he must have leaned two principles: multiplying a number by *n* means adding that number *n* times, and an exponent r represents multiplying the number by itself *r* times. The combination of these principles produces the solution to the problem and the learning of a higher-order principle: To multiply identical variables with exponents, multiply the variable by itself represented by the sum of the two exponents. If a student obtains the right answer to this problem, we usually infer tat he has learned the higher-order principle without requiring him to state it. We confirm our inference by having him solve other problems of the same classes—problems which involve multiplying variables with exponents.

We quote below the views of two authors on the *nature and sequence* of problem-solving activity. One is of Kendler and the other is of Klausmeier. "Problem solving deals with an organism's behaviour in discovering a correct response to a new situation. There are three basic characteristics of problem solving. First, in a problem-solving situation, the correct response is not immediately forthcoming. If it were, there would be no problem. There is a psychological gap between the problem and its solution that can be filled only when the correct response is discovered, Second, human problem-solving beahviour consists largely of mediating responses (or what some prefer to describe as information processing). Research in Problem solving is directed at understanding these mediating events. Factors that Help or Hinder Problem Solving

Motivation is one factor giving direction to people's thoughts in problem-solving. Thinking is quite essential for problem-solving; but soon it shall be clear to you, thinking and problem-solving are not synonymous. What is thinking, then? Earlier it was explained that a symbol is an object or event which stands for, represents, signifies another object or event. Words, statues and national flags too, are instances of symbols. "Thinking can be defined as the mental manipulation of symbols. Of course, we cannot see the symbols being manipulated.

Very often, thinking is impeded by different factors, One such factor is getting trapped in mental ruts. It makes problem solving difficult. Enstellung is a German word which mean mental set. An organism learns a set in a situation and then finds it difficult or impossible to shift to a more appropriate response. In other words, practice in solving problems in one particular sequence tends to give people a "set" to solve similar problems in the same way. "Einstellung' condition entraps them and they forget to be in a "Don't – Be- Blind" condition.

Habit Competition is therefore an important obstacle to Problem-solving. "Problems are often difficult to solve because the incorrect responses persist and thus prevent the correct response. The solution is delayed or even prevented, not because the subject is incapable of making the correct response, but rather because he cannot avoid making the incorrect response. One such phenomenon that underlies the importance of habit competition in problem solving is a "set " or Einstellung, The other is *functional fixedness. Functional fixedness* refers to the difficulty one has in using a familiar object in an unusual manner.

The psychological root of this difficulty is the great strength of the "wrong habit." In other words, functional fixedness is a set that can point thoughts in the wrong direction. It is a set to use objects in the way that we are accustomed to using them, even if a different use might solve the problem. You can understand the meaning of functional fixedness best if we describe a famous demonstration of Maier (1930) which we shall call *The Pliers and the Pendulum Experiment*. The problem is to tie two strings, each of which is hung from the ceiling, together. The strings are so far apart that the subjects cannot reach one string while holding onto the other. The subjects can solve the problem only by tying a weight to the end of one string and converting it into a pendulum, which could be set swinging. When it reaches the top of its swing they can catch it while they hold on to the other string. Scattered about the room are various miscellaneous object including a pair of pliers. The solution is to tie the pliers to one string, to form a pendulum. Swinging the pliers allows one to grasp both strings at the same time. But because of the apparent absence of a weight, few people actually execute the problem. Most fail to realise that the pliers can serve as the weight. They are too used to thinking of pliers only in terms of their primary purpose. In other words, people fix upon the pliers as having the function of a tool and not the function of a pendulum weight. This difficulty is called functional fixedness. It is an aspect of conformity. It keeps us from realising that objects may serve other uses than their conventional ones. "Functional fixedness is one kind of recognition difficulty that depends upon prior experience. Another example involves getting a ping-pong ball out of a tube which is partially blocked at both ends. A variety of tools is provided but none of them can reach the ping-pong ball. The solution is to pour water into the tube to float out the ball.

Groups of people who are given the water in a dirty old bucket solve this problem faster than the groups given the water in an elegant pitcher along with drinking cups and napkins," Problems are often difficult to solve due to habit competition. *Habit Integration* on the other hand, is of much help to problem solving. "Same problems can be solved by combining separate and discrete habits. Two examples of habit integration come from studies designed to examine reasoning and insight". The effect of habit integration in reasoning and problem solving has already been discussed in Chapter nine of this text. An example of habit integration that comes out of studies designed to examine insight is discussed here. Kendler writes:

"A first glance, it seems that insight is not dependent on past experience, that it is unlearned. Such a conclusion, however does not agree with the results of another experiment, in which young chimpanzees were confronted with food just beyond their reach outside the cages. Within reach was a stick they could use to gather in the food. Only one of the six animals solved the problem and he was the one who had previously had extensive experience playing with sticks. The five other animals were then given sticks to play with for a three-day period. During this time, they were never observed using the sticks as rakets; they used them to poke the other chimpanzees or the Experimenter. Yet when these five animals were retested, they solved the problem with ease.

In the previous chapter of this text, we discussed in consider able detail, the *discovery learning* of Bruner and *meaningful verbal learning* of Ausubel. Each of the two approaches has its own merits for providing students with experience in problem-solving. Discovery learning lets the students find things out for themselves. The aim of meaningful verbal learning is to attach readily new information to the existing cognitive structure of the learner. There are times and circumstances when the discovery learning approach is not only acceptable but advisable as it stimulates motivation and creativity. However, the child has a need for security and structure as well as a need to novelty. Therefore, meaningful verbal learning as an organised and logical learning this is related to the individual's prior cognitive structure, has its own utility. It stimulates the thought process of the learner, essential to problem-solving and creativity. If the school has a transmission of knowledge function, it has also the responsibility of developing the student's ability to use acquired knowledge in solving particular problems systematically independently and critical in various fields of enquiry. However, finding problems, as Einstein said, may be important than solving them. Although much study has been done on problem-solving, very little has been devoted to *problem finding*, which involves *creativity*. Encouraging students to identify problems which are of interest to them or which represent needs for them is an excellent approach to give students experience in problem-solving. But it is essential to point out to them that how the problem is stated or defined will determine how it is answered: the narrower the definition, the narrower the solution.

BIBLIOGRAPHY

Annet, J. (1964), *The Role of Knowledge of Results in Learning: A Survey,* Port Washington, New York: U.S. Naval Training Device Centre.

Barker, R.G., Wright, B.A., and Gonick, M.R. (1946), *Adjustment to Physical Handicap and Illness: A survey of the Social Psychology of Physique and Disability,* New York: Social Science Research Council.

Baron, R.A., Byrne, D., and Kantowitz, (1980), *Psychology: Understanding Behaviour,* New York: Holt Rinehart and Winston Inc.

Bekhterev, V.M. (1928), *General Principles of Human Reflexology,* translated by Murphy, G., & Murphy, W., New York, International publishers.

Benedict, R. (1955), *Patterns of Culture,* London: Routlege & Kegan Paul.

Bennett J.W., and Tunin, M. (1948), *Social Life,* New York: Alfred A. Knopf Inc.

Cundill, W., and Weinstein, H. (1969), "Maternal Care and Infant Behaviour in Japan and America." *Psychiatry, 32;* 12-43.

Carmichael, L., ed. (1968), *Manual of Child Psychology,* New Delhi: Wiley Eastern Limited.

Carr, H.A. (1925), *Psychology: A Study of Mental Activity,* New York: David McKay Company Inc.

Cattell, R.B., and Butcher, H. J. (1968), *The Prediction of Achievement and Creativity*, Indianapolis, Indiana: Bobbs Merril.

Doll, E.A. (1948), "The Relation of Social Competence to Social Adjustment," *Educational Record*, XIX.

Doll, E.A. (1953), *Measuring Social Competence: A Manual for the Vineland Social Maturity Scale*, Minneapolis a Education Test Bureau.

Ellis, R.S. (1965), Educational Psychology, New Delhi: Affeliated East-West Press Private Limited.

Eysenck, H.J. (1952), *The Scientific Study of Personality*, London, Routlege & Kegan Paul Ltd.

Eysenck, H.J. (1953), *The Structure of Human Personality*, London: Methuen & Co., Ltd.

Feigenbaum, E.A. (1970), "Information Processing and Memory," In Norman, D.A., ed. *Models of Human Memory*, New York: McGraw-Hill Book Company Inc.

Gagne, R.M. (1964), "Problem Solving," in Melton, A.W., Ed, *Categories of Human Learning*, New York: Academic Press Inc.

Hakes, J.M. (1972), *Child Rearing Practices in North Nigeria*, Ibadan: Ibadan University Press.

Hall, C.S., and Lindzey, G. (1965), *Theories of Personality*, New York: John Wiley & Sons Inc.

Hargreaves, D.H. (1975), *Interpersonal Relations and Education*, London: Routlege & Kegan Paul Ltd.

Joyce, B., and Weil, M. (1985), *Models of Teaching*, New Delhi: Prentice-Hall of India Private Limited.

Judd, C.H. (1937), *The Psychology of Secondary Education*, Boston: Ginn and Company Inc.

Kintisch, W. (1970), "Stochastic Leaning Theory," in Marx M.H., ed. *Learning Theories*, New York: Crowell-Collins-Macmillan Inc.

Klausmeier, J.J. and Ripple, R.E. (1971), *Learning and Human Abilities: Educational Psychology*, New York: Harper & Row Publishers Inc.

Kohler, W. (1925/1927), *The Mentality of Apes*, (translated by E. Winter) New York: Harcourt Brace & World/New York: Humanities Press Inc.

Lazarus, M., Cf. Brown, F.J. (1965), *Educational Sociology*, p. 270, Englewood Cliff, N.J.; Prentice-Hall Inc.

Marx, M.H. ed. (1970), *Learning Theories*, New York: Crowell-Collier-Macmillan Inc.

McDougall, W. (1921), *Introduction to Social Psychology*, Boston: Lace and Company Inc.

McKeller, P. (1957), *Imagination and Thinking*, London: Cohen and West Ltd.

Nihira, K., Foster, R., Shellhan, M., and Leland, H. (1974), *American Association on Mental Deficiency: Adaptive Behaviour Scale*, Washington D.C. : American Association on Mental Deficiency.

Norwell, G.W. (1950), *The Reading Interests of Young People*, Boston: D.C. Health & Company Inc.

Olson, D.R. (1966), "On Conceptual Strategies," in Bruner, J.S., Oliver, R.R. Greenfield, P.M., et al., eds., *Studies in Cognitive Growth*, New York, John Wiley & Sons Inc.

Owen, S.V., Blount, H.P., and Moscow H. (1978), *Educational Psychology: An Introduction*, Boton: Little, Brown & Company Inc.

Page, J.D. (1947), *Abnormal Psychology*, New York: McGraw-Hill Book Company Inc.

Parkhurst, H. (1922), *Education on the Dalton Plan*, New York: E.P. Dutton and Company Inc.

Reitman, J.S. (1970), "Computer Simulation of an Information Processing Model of Short-Term Memory," In Norman, A., ed., *Models of Human Memory*, New York, Academic Press Inc.

Schank, R.C., and Colby, K.M. eds. (1973), *Computer Models of Thought and Language*, Sanfransisco: Freeman.

Seidman, J.M., ed., (1960), *The Adolescent: A Book of Readings*, New York: Holt, Rinehart and Winston Inc.

Sharpe, E.F. (1950), *Collected Papers on Psychoanalysis*, London: Hogarth Press Ltd.

Taba, H. (1966), "Teaching Strategies and Cognitive Functioning in Elementary School Children," *Cognitive Research Project*, 2404, Sanfransisco: Sanfransisco Stare College.

Thomspon, H. (1968), "Physical Growth," In Carmichael, L., ed., *Manual of Child Psychology*, pp. 292-334, New Delhi: Wiley Eastern Private Limited.

Turner, R.H., and Killian, L.M. (1957), *Collective Behaviour*, Engewood Cliffs, N.J.: Prentice-Hall Inc.

Tyler, L.E. (1968), *Psychology of Human Differences*, Bombay: Vakil, Feffer & Simons, Pvt., Ltd.

Underwood, B.J. (1957) "Interference and Forgetting," *Psychological Review*, 64: 49-60.

Wallas, G. (1926), *The Art of Thinking*, New York: Harcourt, Brace & World, Inc.

INDEX

***